ACPL IT DISCARDED

S0-BUX-352

Flexible manufacturing islands

**DO NOT REMOVE
CARDS FROM POCKET**

**ALLEN COUNTY PUBLIC LIBRARY
FORT WAYNE, INDIANA 46802**

You may return this book to any agency, branch,
or bookmobile of the Allen County Public Library.

DEMCO

GARLAND STUDIES ON INDUSTRIAL PRODUCTIVITY

edited by
STUART BRUCHEY
UNIVERSITY OF MAINE

A GARLAND SERIES

FLEXIBLE MANUFACTURING ISLANDS

STEPPING STONES IN FMS IMPLEMENTATION

MEHDI KAIGHOBADI

GARLAND PUBLISHING, INC.
NEW YORK & LONDON / 1994

Allen County Public Library
900 Webster Street
PO Box 2270
Fort Wayne, IN 46801-2270

Copyright © 1994 by Mehdi Kaighobadi
All rights reserved

Library of Congress Cataloging-in-Publication Data

Kaighobadi, Mehdi.
　Flexible manufacturing islands : stepping stones in FMS implementation / Mehdi Kaighobadi.
　　p.　cm. — (Garland studies on industrial productivity)
　Includes bibliographical references and index.
　ISBN 0-8153-1612-7
　1. Flexible manufacturing systems.　2. Production planning.
I. Title.　II. Series.
TS155.6.K35　1994
670.42'7—dc20　　　　　　　　　　　　　　　　93-47243
　　　　　　　　　　　　　　　　　　　　　　　　　CIP

Printed on acid-free, 250-year-life paper
Manufactured in the United States of America

This work is dedicated to my wife, Joni, and my son, Kaivan, who tolerated long hours of research leading to this work and whose support made the job less demanding than it could have been.

TABLE OF CONTENTS

CHAPTER I: *INTRODUCTION*

    Statement of the Problem 4
    Purpose and Objectives of the Study 5
    Scope and Limitations of the Study 7

CHAPTER II: *REVIEW OF LITERATURE*

    Benefits of FMS 11
    Problems Associated With FMS 12
    Analytical Models of FMS 14
    Lot Sizing and Scheduling in the FMS Environment ... 17
    Lot Sizing and Scheduling in Conventional Systems ... 20
    Item Sequencing Considerations in Scheduling 24

CHAPTER III: *RESEARCH DESIGN AND METHODOLOGY*

    The Choice of Methodology 28
    Definitions 29
    Assumptions and Description of the Model 33
    Detailed Procedure of the Study 34
    Conventional System 36
    Common Work Center Converted to FMI 50
    Non-Common Work Center Converted to FMI 54

CHAPTER IV: *RESULTS*

    Configuration I 62
    System Overview 62
    Case 1: Conventional System 64
    Case 2: Work Center B is FMI 69
    Case 3: Work Center C4 is FMI 72

    Configuration III 79
    System Overview 79

    Case 1: Conventional System 82
    Case 2: Work Center B is FMI 86
    Case 3: Work Center A4 is FMI 91

    Configuration II 100
    System Overview 100
    Case 1: Conventional System 101
    Case 2: Work Center B is FMI 109
    Case 3: Work Center C2 is FMI 114
    Case 4: Work Center A2 is FMI 119

CHAPTER V: *CONCLUSIONS*

    Conclusions 128
    Suggestions for Further Research 131

APPENDIX 132
REFERENCES 137
INDEX 147

LIST OF TABLES

Table	Description	Page
1	Comparison Conventional and FMS Systems	13
2	Work Center Candidates for Conversion to FMI	32
3	Results Matrix	32
4	Parameters of the Simplified System	35
5	Annualized Total Inventory Holding Cost for Various Items Sequences	43
6	List of Unique Alternating Items Sequences	47
7	Annual Demand Rates for Sensitivity Analysis	49
8	Set-up Times for Sensitivity Analysis	49
9	Inventory Holding Costs for Sensitivity Analysis	50
10	Inventory Holding Costs after Bottleneck Work Center is Converted into FMI	52
11	Annual Inventory Holding Costs after Center A1 Has been Converted into FMI	56
12	Summary of Simplified Example Results for Item Sequence 1-2-3-4	59
13	System Parameters for Configuration I	63
14	Production Schedule and Inventory Behavior Following Work Center B: Conventional System, Base Case, Sequence A	64

Table	Description	Page
15	Inventory Behavior Following Work Center C1: Configuration I, Conventional System, Base Case, Sequence A	65
16	Annual Inventory Holding Cost: Configuration I, Conventional System, Base Case, Sequence C or E	68
17	Total Inventory Holding Cost Under Sensitivity Analysis: Configuration I, Conventional System, Sequence C or E	68
18	Comparison of Inventory Holding Costs Under the Base Case and the Sensitivity Analyses: Configuration I, Conventional System, Sequence C or E Sensitivity Analysis	69
19	System Inventory Holding Costs: Configuration I, B is FMI, Base Case, Sequences C or E	71
20	Inventory Holding Costs Under Sensitivity Analysis: Configuration I, B is FMI, Sequence C or E	71
21	Comparison of Inventory Holding Cost Under the Base Case and the Sensitivity Analyses: Comparison I, B is FMI, Sequence C or E	72
22	Annual Inventory Holding Cost: Configuration I, Base Case, C4 is FMI, Sequence C	74
23	Annual Inventory Holding Cost Under Sensitivity Analysis: Configuration I, C4 is FMI, Sequence C	74

Table	Description	Page
24	Comparison of Inventory Holding Cost Under Base Case and Sensitivity Analysis: Configuration I, C4 is FMI, Sequence C	75
25	Inventory Holding Cost Under Various Cases and FMI Positions: Configuration I Positions	76
26	System Parameters for Configuration III	81
27	Production Schedule and Inventory Behavior Following Work Center B: Configuration III, Conventional System, Base Case, Sequence A	82
28	Inventory Behavior Following Work Center A1: Configuration III, Conventional System, Base Case, Sequence A	83
29	Annual Inventory Holding Cost: Configuration III, Conventional System, Base Case, All Five Sequences	84
30	Total Inventory Holding Cost Under Sensitivity Analyses: Configuration III, Conventional System, Sequence C or E	85
31	Comparison of Inventory Holding Cost Under the Base Case and the Sensitivity Analysis: Configuration III, Conventional System, Sequence C or E	85
32	Inventory Behavior Following Work Center A1: Configuration III, B is FMI, Base Case, All Sequences	87

Table	Description	Page
33	System Inventory Holding Cost: Configuration III, B is FMl, Base Case, All Sequences	89
34	Inventory Holding Cost Under Sensitivity Analyses: Configuration III, B is FMI, Sequences C or E	89
35	Comparison of Inventory Holding Cost Under the Base Case and the Sensitivity Analyses: Configuration III, B is FMI, All Sequences	90
36	Inventory Behavior Following Work Center A4: Configuration III, A4 is FMI, Base Case, Sequence A	92
37	Annual Inventory Holding Cost: Configuration III, Base Case, A4 is FMI, All Sequences	93
38	Annual Inventory Holding Cost Under Sensitivity Analyses: Configuration III, A4 is FMI, All Sequences	94
39	Comparison of Inventory Holding Cost Under the Base Case and the Sensitivity Analyses: Configuration III, A4 is FMI, All Sequences	94
40	Inventory Holding Cost Under Various Cases & FMI Positions: Configuration III	95
41	System Parameters for Configuration II	102

Table	Description	Page
42	Production Schedule and Inventory Behavior Following Work Center B: Configuration II, Conventional System, Base Case, Sequence A	103
43	Production Cycle for Work Center A1: Configuration II, Conventional System, Base Case, Sequence A	104
44	Production Cycle for Work Center C1: Configuration II, Conventional System, Base Case, Sequence A	105
45	Inventory Behavior Following Work Center A1: Configuration II, Base Case, Conventional System, Sequence A	106
46	Inventory Behavior Following Work Center C1: Configuration II, Base Case, Conventional System, Sequence A	106
47	Inventory Holding Cost: Configuration II, Conventional System, Base Case, All Five Sequences	106
48	Annual Inventory Holding Cost: Configuration II, Conventional System, Base Case, Sequence C	107
49	Total Inventory Holding Cost Under Sensitivity Analyses: Configuration II, Conventional System, Sequence C	108
50	Comparison of Inventory Holding Cost Under the Base and the Sensitivity Analyses: Configuration II, Conventional System, Sequence C	108

Table	Description	Page
51	System Inventory Holding Cost: Configuration II, B is FMI, Base Case, Sequence A	111
52	Inventory Holding Cost Under Sensitivity Analyses: Configuration II, B is FMI, Sequence A and D	113
53	Comparison of Inventory Holding Cost Under the Base Case and The Sensitivity Analysis: Configuration II, Base Case, B is FMI, Sequence C	113
54	Inventory Holding Cost: Configuration II, Base Case, C 2 is FMI, All Five Sequences	115
55	Annual Inventory Holding Cost: Configuration II, C2 is FMI, Base Case, Sequence E	116
56	Inventory Holding Cost Under Sensitivity Analyses: Configuration II, C2 is FMI, Sequence E	117
57	Comparison of Inventory Holding Cost Under the Base Case and the Sensitivity Analyses: Configuration II, C2 is FMI, Sequence E	118
58	Annual Inventory Holding Cost: Configuration II, A2 is FMI, Base Case, Sequence C	120
59	Inventory Holding Cost Under Sensitivity Analyses: Configuration II, A2 is FMI, Sequence C	121

Table	Description	Page
60	Comparison of Inventory Holding Cost Under the Base Case and the Sensitivity Analyses: Configuration II, A2 is FMI, Sequence C	121
61	Inventory Holding Cost Under Various Cases and FMI Positions: Configuration II	122
62	System Total Annual Inventory Holding Cost: Various Configurations and FMI Positions	126
63	Percentage Reduction in Inventory Holding Cost of the System Under Various Configurations and FMI Positions	127
64	Annual Demand Under Sensitivity Analysis: Configuration I	132
65	Setup Times Under Sensitivity Analysis: Configuration I	132
66	Inventory Holding Cost/Unit Under Sensitivity Analysis: Configuration I	133
67	Annual Demand Under Sensitivity Analysis: Configuration II	134
68	Setup Times Under Sensitivity Analysis: Configuration II	134
69	Inventory Holding Cost/Unit Under Sensitivity Analysis: Configuration II	135
70	Annual Demand Under Sensitivity Analysis: Configuration III	135
71	Setup Times Under Sensitivity Analysis: Configuration III	136

Table	Description	Page
72	Inventory Holding Cost/Unit Under Sensitivity Analysis: Configuration III	136

LIST OF ILLUSTRATIONS

Table	Description	Page
1	Configuration of the System in the Research	10
2	A Conceptual Design of Flexible Manufacturing System	30
3	Inventory Behavior Following Work Center B	39
4	Scheduling Chart of Work Centers B, A1, and A2	40
5	Inventory Behavior Following Work Center A1	45
6	Inventory Behavior Following Work Center A2	46
7	Inventory Behavior Following Work Center A1 When Work Center B has Been Converted into FMI	53
8	Inventory Behavior Following Work Center A2 When Work Center B Has Been Converted	54
9	Inventory Behavior Following Work Center A1, When Work Center A1 Has Been Converted into FMI	57
10	Comparison of Annualized Inventory Holding Costs of Totally Conventional System and the System With an FMI in Various Positions	58
11	System Schematic for Configuration I	62
12	Comparison of Inventory Holding Cost: Capacity Utilization Versus FMI Position, Configuration I	77

Table	Description	Page
13	Comparison of Inventory Holding Cost: Setup Time to Run Time Ratio, Versus FMI Position, Configuration I	77
14	Comparison of Inventory Holding Cost: Rate of Increase in Inventory Holding Cost Versus FMI Position, Configuration I	78
15	System Schematic for Configuration III	79
16	Comparison of Inventory Holding Cost: Capacity Utilization Versus FMI Position, Configuration III	98
17	Comparison of Inventory Holding Cost: Setup Time to Run Time Ratio Versus FMI Position, Configuration III	98
18	Comparison of Inventory Holding Cost: Rate of Increase in Inventory Holding Cost Versus FMI Position, Configuration III	99
19	System Schematic for Configuration II	100
20	Comparison of Inventory Holding Cost: Capacity Utilization Versus FMI Position, Configuration II	123
21	Comparison of Inventory Holding Cost: Setup Time to Run Time Ratio Versus FMI Position, Configuration II	123
22	Comparison of Inventory Holding Cost: Rate of Increase in Inventory Holding Cost Versus FMI Position, Configuration II	124

PREFACE

In recent years the potential of factory automation in general and Flexible Manufacturing Systems (FMS) in particular has been the subject of numerous studies. Along with the success cases, a number of investigations point to problems associated with the introduction of various types of factory automation, particularly FMS. Research indicates that when the introduction of such technology is incremental, the success of the implementation is often hindered by lack of integration of the new technology with the existing conventional system. The purpose of the present research is to investigate operational issues associated with the introduction of a flexible manufacturing work center (referred to as a Flexible Manufacturing Island or FMI) into an otherwise conventional system. The primary objective of the research is to investigate the impact on system performance associated with the installation of a flexible manufacturing island in an otherwise conventional manufacturing system. More specifically, the study will compare the relative impacts on system inventory holding cost caused by installation of a single FMI at different locations within the system. Finally, the study will investigate the sensitivity of these impacts on inventory holding cost to changes in (1) system utilization, (2) the ratios of setup times to run times in the conventional work centers, and (3) the rates of increase in holding costs for items as they move through the system.

This relatively complex research methodology and flow is explained by using a simplified example. The system under the present study has five work centers and three configurations of these work centers will be studied. The methodology employed is simulation in that system inventory behavior is projected through time. Methods are developed or identified for determining various factors of the study, such as optimum production cycle duration and near optimum item sequences. The results of the simplified example for the totally conventional system and for two alternative locations for the FMI are shown and discussed. It should be noted that in presenting the three configurations of system for the study, Configuration III is discussed before II. This is because the complexity of Configuration II requires that some fundamental relationships be explained in the simpler system of Configuration III.

The intention of the author is to shed some light on the complex issues of decision making in the arena of factory automation and help understand the interaction of a major strategic decision (replacing a conventional system with an automated one) on the short-term tactical aspect of manufacturing operations (in this case, inventory holding cost). Hopefully, this research will open avenues for further questions and research.

Mehdi Kaighobadi

November 1993, Florida

ACKNOWLEDGEMENT

Many people have contributed to this work and without their continuous support completion of this research would have been a lot more difficult. I wish to extend my sincere thanks and gratitude to those who contributed their encouragement, advice, support, and patience to the finalization of this work.

First and foremost, I am deeply indebted to Dr. Thomas B. Clark of Georgia State University for his ever-present support, advice, and guidance throughout the process of developing the research idea, conducting the research analysis, and its completion. I wish also to thank Dr. Richard Deane and Dr. Bikramjiat S. Garcha, both from Georgia State University, for their helpful suggestions toward the completion of this work. I also would like to extend my appreciation to Mrs. Pat Kennedy, our secretary, at Florida Atlantic University, who patiently contributed to the word-processing of this document.

Last, but not least, I would like to extend my sincere appreciation to my wife for her complete support. Joni has put up with a lot during this research.

Flexible Manufacturing Islands

CHAPTER I

INTRODUCTION

Global market pressure has been a primary driving force behind recent advances in the manufacturing technologies in several industries. Various solutions and prescriptions have been proposed for American manufacturers to regain their once envied status in the global market. Factory automation is at the top of the list of such prescriptions. Many researchers and practitioners have claimed that factory automation, in one form or another, will resolve many of the problems faced by U.S. manufacturers. For instance, Hughes and Hegland [1] propose that the traditional trade-off between low-volume production with flexibility on the one hand and high-volume production with low unit-cost on the other can be resolved by utilizing Flexible Manufacturing Systems (FMS). In a broader sense, Computer Integrated Manufacturing (CIM) has been hailed as the key to gaining manufacturing competitiveness [2, 3, and 4]. The range of benefits derived from advanced technologies are intended to help a firm become more competitive by being both more efficient and more flexible in responding to changes in the marketplace.

Benefits claimed to ensue from FMS cover a wide range [5, 11], including:

- Reduced setup time
- Reduced inventory cost
- Reduced throughput time
- Improved product quality
- Increased flexibility in responding to market changes, both in the variety of product demanded and in the variation of the level of demand.

Zisk [6] believes that "flexible manufacturing is a practical, affordable, and effective strategy for remaining competitive." Other benefits, such as reduction in floor space, decreased need for different specialized machine tools, and increased machine utilization have also been reported [7]. Hartley [10] describes these benefits in detail, citing actual companies with FMS. Professional journals have devoted entire special issues to case studies describing the benefits of FMS and the "success" stories of the firms installing such systems [12]. The advantages of automated manufacturing, including FMS, have even been recognized by the general media [8]. It should be noted that FMS does not render the same level of benefits for different types of manufacturing industries. The advantages of FMS are substantial in industries involving the machining of parts. A research study by Cincinnati Milacron, a major supplier of FMS technology, shows that a typical work piece in the machining industry spends only about 30 percent of the working time actually on the machines. The rest of the time is used to setup machines, transfer the work piece from one work station to another, sort tools, and so on [9]. The advantages of FMS are fully realized in systems similar to the ones studied by Cincinnati Milacron, because FMS will drastically reduce the time the work piece is idle. Of course, the benefits of FMS are more realizable when the flexible technology is fully implemented and integrated.

Statement of the Problem

Despite all these "claims" of benefits, FMS has failed in many cases by one measure or another. These failures have been found both by researchers involved in factory automation [13] and by firms that have already installed such systems, such as the John Deere Company and General Motors [16]. Lack of integration seems to be at the heart of such failures. Integration of the components of a manufacturing system is vitally important for achieving high performance. Mize and Seifiet [14] emphasize the need for integration in a firm's overall manufacturing structure and believe that creating this integration is the biggest challenge for system designers.

The effect of non-integrated systems is often observed when the technological development of a factory follows an incremental path, from a completely conventional manual system, to "islands of automation," to a Flexible Manufacturing

System, and ultimately to a Computer Integrated Manufacturing System (CIMS). In fact, some researchers believe that this incremental approach is the appropriate strategy for firms that do not have the resources and/or readiness required to implement a full-fledged FMS all at once [15]. The current trend in industry is to start by installing a few flexible manufacturing cells (called flexible manufacturing islands or FMI in this study) and gradually build a complete FMS [17]. This approach creates the potential for a situation which is well described by Church [72]:

> *Except in rare instances, it is safe to say that flexible manufacturing will be introduced by degrees. This piece by piece or piecemeal introduction is the source of a potential problem. Inadequate attention may be given to the interface of the flexible manufacturing subsystems to the remaining conventional manufacturing systems.... The full potential of the flexible manufacturing subsystems may not be reached because of constraints imposed by the manufacturing subsystems which supply the flexible subsystem and which use the product produced. The effectiveness of the conventional subsystems may also be impaired by introduction of flexible manufacturing subsystems....There is a real danger that flexible manufacturing subsystems will become ineffective islands of automation.*

Purpose and Objectives of the Study

The general purpose of this research is to investigate operational issues involved in the incremental installation of FMIs in otherwise conventional manufacturing systems. Wemmerlov and Hyer [32], in their article on the research issues in cellular manufacturing, indicate that research addressing such issues can provide vitally important guidance for potential users of such advanced manufacturing technologies.

The primary objective of this study is to investigate the impact on system performance associated with the installation of a flexible manufacturing island in an otherwise conventional manufacturing system. More specifically, this study will compare the relative impacts on system inventory holding cost caused by installation of a single FMI at different locations within the system. Finally, the study will investigate the sensitivity of these impacts on inventory holding cost to changes in: (1) system utilization, (2) the ratios of setup times to run times in the conventional work centers, and (3) the rates of increase in holding costs for items as they move through the system.

In this research, the term "flexible manufacturing island" is used to distinguish the object under study from completely integrated Flexible Manufacturing Systems. In an FMS, usually all or a majority of components are automated (i.e., machining centers, transportation mechanisms, and computer control system). In the system to be studied here, only one of the work centers in the system is changed from conventional to a flexible work center. For the purposes of this study, the operational characteristic that will distinguish an FMI from a conventional work center is that setup times required to switch from the production of one item to another are negligible in an FMI, whereas the setup times will be significant in the conventional work centers. This characteristic also introduces the possibility of item-by-item material flows to and/or from an FMI versus batch flows typically observed among conventional work centers.

The type of manufacturing system to be addressed in this study is a "closed hybrid-flow shop." The distinguishing characteristics of this type of system are:

Chapter 1: Introduction 7

1. The system produces a set of standard items, each with constant demand. These standard items are produced on a repetitive cyclical basis (thus "closed").
2. Different items follow different paths through the system, but all paths flow in the same general direction through the shop (thus "hybrid-flow"). More precisely, all item flows between any two work centers are in the same direction.

Scope and Limitations of the Study

This investigation will address issues of integration in mixed conventional/flexible manufacturing systems in which flexible work centers are being installed incrementally. This study does not address full-fledged FMSs. Many studies have focused on the total FMS environment, and such studies will be reviewed in Chapter II. This study will not consider the economic justification of an FMI, nor will it study the direct financial impact of replacing conventional work centers with FMIs, except for the impacts on inventory holding cost. Again, a number of studies have addressed these aspects of factory automation, and a brief review of such studies will be presented in chapter two. More specifically, the following paragraphs define the scope and limitations of this research:

1. The manufacturing system under investigation is a hypothetical system with the distinguishing characteristics of a "closed hybrid-flow shop" as described previously. This system does not represent an existing real-world manufacturing system. However, the characteristics of the hypothetical system enable the researcher to address the issues involved in the study under controlled conditions.
2. Of the many possible configurations for the type of system studied here, three configurations will be investigated. These three configurations are illustrated in Figure 1. These configurations have the following characteristics:

a. Each configuration contains one work center that is common to the paths of all items. This "common" work center may be at the beginning (Configuration I), in the middle (Configuration II), or at the end (Configuration III) of the system.
b. Each configuration contains five work centers. This is the minimum number of work centers required to have item flows converge before and diverge after the common work center in Configuration II. All other work centers will be referred to as "non-common" work centers in this study.

Real world systems with similar flow characteristics can be viewed as combinations of the three basic configuration under investigation in this study.

3. No attempt is made to identify or model the specific operations performed at individual work centers. In other words, "generic" work centers are assumed in this study.
4. The system is assumed to produce eight items. This number is the minimum necessary for multiple (i.e. exactly 2) items per path. Thus, at least two items will flow through any given work center.
5. Comparisons between conventional and FMI systems are made under near-optimum operating disciplines with respect to the impact on inventory holding cost. Specifically, each system will be evaluated under (a) an optimum production cycle duration, which determines the lot sizes for each item, and (b) a near optimum sequence of items through the common work center. The use of "near-optimum" item sequences is due to the difficulty of finding and verifying the absolute optimum item sequence.
6. This study considers the impact of introducing FMIs on inventory holding cost only. No attempt is made to deal with other financial impacts of FMIs.

Chapter 1: Introduction 9

7. This study does not investigate the impacts of machine breakdowns or quality problems on the performance of the systems under study.

In Brief

In this chapter, the problem which prompted this study was presented. The objective of the study was specified to be investigating the impact on inventory holding cost of introducing a Flexible Manufacturing Island (FMI) into an otherwise conventional system. Finally, the scope of the research was discussed, and its limitations were identified.

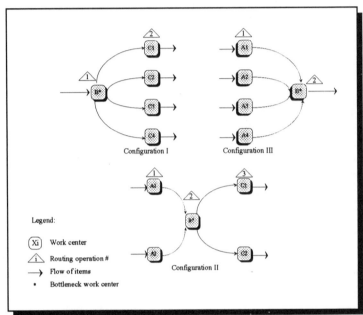

Figure 1. Configurations of the System in the Research

CHAPTER II

REVIEW OF LITERATURE

Although Flexible Manufacturing Systems are in an early stage of development, the literature on various aspects of such technological advances is growing in scope and depth very rapidly. In addition, there is an enormous body of literature dealing with scheduling, design, and inventory control of conventional systems. Both of these areas of literature are relevant to this research. The review of literature will be limited to those works which specifically relate to this study.

Benefits of FMS

A vast number of case studies and reports have described the benefits of automated factories in general and Flexible Manufacturing Systems in particular. A brief review of this literature is presented here. Most of the reports about the benefits of computerized factories appear in trade and professional journals such as *Industrial Engineering, Modern Material Handling, American Machinist, Production, Production Engineering,* and *Managing Automation.* Some case studies are considered so important that special issues of various journals have been devoted to covering them. For instance, *Industrial Engineering* devoted a two-year series to the coverage of Computer Integrated Manufacturing Systems, and a large number of the articles appearing in the series dealt with case studies of firms with FMS [17]. A major survey on factory automation was conducted by The Office of Technology Assessment [18]. In this study, such issues as the impact of various forms of automation, collectively called "Programmable Automation," on human resources, level of worker displacement, and overall work environment were addressed. An interesting study was done by Solomon and Beigel [19] in which they compared conventional manufacturing systems with FMS from various points of view and under different probability states of FMS performance. Table 1 summarizes the results of this study. The authors indicate that small as well as large manufacturers can benefit from FMS. In fact, this conclusion is one justification for the present study, since small manufacturers typically cannot afford to invest in large scale factory automation and usually begin with small, incremental, and progressive conversion of their conventional manufacturing systems to FMS. Various researchers have studied actual Flexible Manufacturing Systems in operation throughout the U.S. [see, for example, 20, 21, 22, 23, 36, and 72]. Most of these studies focus on the benefits realized by the firms installing such systems. A general conclusion which could be drawn from these case studies is that FMS has a wide range of potential benefits, including:

- Saving space
- Allowing better tool control
- Providing flexibility in meeting customers needs
- Reducing work-in-process
- Reducing parts shortage
- Improving product quality
- Increasing employee morale
- Increasing quality of work life

Problems Associated With FMS

Despite all the positive reports on the benefits of FMS and similar computerized manufacturing technologies, a number of reports and studies also point to the problems associated with such systems and the factors which hinder attainment of their potential. As Harvey [24] indicates, the main barrier to successful installation and implementation of FMS is the problem of integrating individual parts of the factory, and accomplishing this goal is highly dependent upon a well-designed plan of FMS integration within an existing system.

Suri and Whitney [25] point out that while a number of FMSs have been installed by various firms aiming at increasing their productivity, these systems have been mostly under-utilized. The authors go on to indicate that the main reason for such under- utilization is the complexity of these advanced technologies and the lack of adequate control support systems. Suri and Hildebrant [26] also point to the complexity of FMS and state that development of analytical and simulation models, such as the one developed by the authors themselves (known as Mean-Value-Analysis-Que, or MVAQ) would be very helpful in predicting the performance of such systems.

Table 1
Comparison of How Machines and Work Parts Spend Their Time in the Shop of conventional System and in FMS Under Various Levels of FMS Performance

PARAMETER	CS	MP	ML	MO
Percent of machine time the machine spends without parts	50	35	20	5
Percent of machine time that there is a part on the machine	50	65	80	95
Percent of time that part is not being worked on while on the machine	70	35	21	7
Percent of time that the part is being worked on while on the machine	30	65	79	93
Percent of manufacturing lead time part spends time that the either moving or waiting	95	92.5	90	85
Percent of manufacturing lead time that the part spends on the machine	5	7.5	10	15

CS = Conventional Systems   MP = Most Pessimistic;
ML = Most Likely   MO = Most Optimistic
Source: Solomon and Biegel [19]

The need for such models is increased when components of an FMS are to be installed in a conventional system.

Quoting Professor Stecke, an expert in the area of FMS, Piszczalski [27] states the problem of integration as

> .....*the flexible system may quickly feed parts to an older system that can't match the throughput rates of the [flexible system] or requires large batches at a time and so can't handle a quickly changing part mix. Therefore a balance of flexibility among various stages of production is important.*

Similarly, Ferrario [28] points to the problem of integration of FMS into an existing manufacturing system and states that "The subject of flexible manufacturing systems is one which strikes joy into the heart of the advanced manufacturing technology engineer and terror into the heart of the manufacturing engineer. The reason for both these effects is integration."

Analytical Models of FMS

Another vast body of literature relates to the models developed to address various issues relating to factory automation. These studies generally provide frameworks, heuristics, or mathematical models which describe the behavior of FMS and/or provide rules for planning and control of such systems.

One of the major contributions to the literature on FMS models is attributable to Buzacott and Yao [24]. They present a comprehensive review of FMS analytical models. The authors first classify these models into "Purdue models", "MIT models", "Draper Labs", "Toronto" and "others", depending on who were the original developers of the models. The authors also indicate that "the dominant modeling tool in studying FMS has been simulation." The classification and review provided in the study by Buzacott and Yao give insight into the

Chapter 2: Literature Review 15

trend in the development of analytical models as they relate to FMS and shows that the main issues in this study have not yet been addressed. Similarly, Wilhelm and Sarin [30] review the mathematical models that focus on planning and operational issues of FMS. Their work is of the same nature as the study by Buzacott and Yao [24]. However, they provide more details for the justification of various research and design tools in planning for FMS. For instance, they indicate that queuing network models have been mostly applied to and are appropriate for planning, rather than operating problems of FMS, while simulation can be applied both to planning and operating problems associated with FMS design.

Valcada and Masttretta [29] compare two major classes of models, (simulation versus analytical models) and present pros and cons of each set of methodologies. The main conclusion that the authors draw from their study is that in the early stage of planning for FMS design, analytical and simple simulation models are more helpful, whereas after the system's configuration has been well defined, detailed simulators provide better design tools. This conclusion supports the choice of simple simulation together with an analytical model for this study, since the study will address issues commonly encountered during the early phases of planning for FMS.

One of the earlier studies on FMS was performed by Buzacott and Shantikumar [30] in which the authors specified the major characteristics of FMS and developed a model for determining the production capacity of such systems. This study is relevant to this research in that it provides the basic conceptual framework of the system being studied here and provides useful insights into the characteristics of such systems. For instance, the authors indicate that the sequence and timing for releasing jobs to the system as well as the movement of parts inside the system, must be determined both at the planning and at the control level of systems design. Similarly, a general model developed as a decision support system for FMS by Suri and Whitney [33] provides useful insights into the basic operating parameters of such systems.

Maimon [34], using a hierarchial approach, develops a general FMS model. The author provides a detailed control scheme for FMS based on the various levels of decision making involved in such an environment. His study focuses on short term issues of controlling an FMS and provides a helpful framework for addressing operational decisions in such systems. These issues range from the production requirement parameters to the actual control of part movements among the work centers. Maimon's study provides the insight needed to address short term control problems of FMS which would be prevalent when an FMI is introduced into a conventional system.

A number of studies have focused on the issues and problems of designing computerized manufacturing systems. In the context of cellular manufacturing, Flynn and Jacobs [37] compare process layout and cellular layout design with respect to various performance criteria, including average setup time, average number of parts completed, average waiting time, and average work-in-process inventory. However, their study focuses on group technology and cellular manufacturing and does not address the issue of a hybrid system, where FMS is integrated into a conventional system. Another study which is close to this research was conducted by Burstein [38]. The main contribution of the study is a set of decision rules to assist in the introduction of flexible and/or dedicated subsystems into manufacturing systems. Although the author addresses questions similar to those involved in this study, the system being investigated in his study is a transfer line system into which a full FMS (not an FMI) is introduced. The study does not address the hybrid flow shop considered in this study. Similarly, a study by Costa and Garetti [39] focuses on developing a design for control of manufacturing cells. Although this study addresses the job-shop environment, it assumes pure computerized flexible manufacturing machines and does not address the conventional component of this study. A similar study by Cutosky, Fussel, and Milligan [40] developed a design model for small batch producing flexible machining cells. The model developed a machining cell design that, according to the authors, "is easily integrated into a larger system and is readily modified or expanded as more sophisticated equipment and techniques become available." However, the model is intended to help integrate a flexible machining cell into a broader FMS or CIMS and does not address the problem of introducing an FMI into a conventional production system. Hall and Stecke [41] address the design problem of computerized manufacturing systems. However, they focus only on flexible assembly systems (FAS). Furthermore, Hall and Stecke's study does not compare any two systems on the basis of operating performance. The study addresses only general issues in designing FAS. Various other studies have addressed different design issues in CIM or in FMS, ranging from technical aspects of such designs [42], to operational problems of such systems (such as part type selection, tooling, fixture allocation, etc.) [43], to the physical layout of FMS (and how they are linked to the material handling systems) [44], to the application of "Computer Integrated Architecture" to FMS [45]. Addressing design issues relates all the models reviewed in this section to this research. However, what distinguishes these studies from the objectives of this research is that none of the studies reviewed here address the integration of an FMI into an existing conventional system. The primary contribution of the research reviewed so far to this study is that they point to the potential issues that may arise in the methodology of the present study.

Lot Sizing and Scheduling in the FMS Environment

Questions relating to lot sizing and scheduling are important issues in this study. Therefore, it is fruitful to review the literature with respect to the studies done in these areas. If one were to divide a mixed conventional/FMI system (like the one to be studied in this research) into two segments, they would be the conventional subsystem and the flexible manufacturing subsystem. In the preceding section, the literature on design issues for the flexible segment was reviewed. In what follows, some of the major studies focusing on lot sizing and scheduling of FMS will be reviewed. Next the

Flexible Manufacturing Islands 18

literature on these same issues in conventional systems will be analyzed, and their relationships/contributions to the present study will be highlighted. The focus will be on those studies that address multi-stage, multi-product systems, since this is the type of system to be investigated in this study.

Lot sizing in a Flexible Assembly System (FAS) is the subject of a study by Chang and Sullivan [46]. In their model, FAS periodically produces subassemblies to support main assembly line(s). The objective of the study was to develop a framework for determining lot sizes in such an environment in order to optimize several criteria. The objective function of the model was presented in the form of a cost function. The major differences between this study and this research are that (a) the model does not allow for a conventional subsystem and (b) it is restricted to an assembly flowline with a fixed cycle time that applies to all work stations along the line. The framework presented by Chang and Sullivan can contribute to selecting near-optimal operating disciplines for the various system configurations to be investigated.

Chakravarty and Schtub [47] address the impact of the flexibilities offered by FMS on operational planning, such as Master Production Scheduling (MPS). The authors focus on planning issues in a pure FMS environment, and they do not directly investigate such issues in a mixed conventional/FMI system.

Hildebrant [69] has studied the problem of scheduling in Flexible Machining Systems. Flexible Machining Systems form a subclassification of Flexible Manufacturing Systems, and specifically they include typical job-shop oriented machine shops. Apart from this difference, the two terms can be used interchangeably. Hildebrant develops an approximation model for allocation of work to machines and the input strategy for such systems. The approximation approach is used because such systems have complicated structures which would make exact optimization approaches very computationally demanding. Hildebrant's study only addresses fully implemented and integrated flexible systems and the problem of work assignment. The issues of installing an FMI into a conventional system are not investigated.

In the same vein, Afentakis [48] developed a model which maximizes the throughput of an FMS by solving the machine loading problem and part sequencing problem. In the model, the author maximizes the production rate by minimizing the cycle time which is defined as "the time required to produce one unit of the final assembly, i.e. one unit of each part [type]." (p. 613). This portion of the model is relevant to the present study, since this study also attempts to find the minimum feasible cycle time in order to minimize inventory holding costs. However, it should be noted that the model developed by Afentakis does not consider various setup times for different work stations, since it addresses FMS only where setup times at each machine are considered near-zero and therefore negligible.

Comparison of Group Technology (GT) in the form of Cellular Manufacturing and FMS is the subject of a study by Vaithianathan [49]. However, his study is different from this research in that the existence of group cellular operations, which "is often dedicated to the production of one family of parts" (p. 421) is not assumed in this study.

Mabert and Pinto [56], in their study of one of Ford Motor Company's fabrication plants, investigated the impacts of various demand rates, number of setups, parts costs, process times, and available capacity changes on the behavior of a Flexible Assembly System (FAS). In this study they recognize the "sunk" nature of direct labor cost in the short run in an automated system. As far as this research addresses similar variables (such as holding costs, setup times, etc.), the study by Mabert and Pinto is related to this study. The main contribution of the Mabert and Pinto study to this research is the set of relationships that the authors describe among processing time, periodic requirement for parts, setup time for parts, and production rate. These relationships led to the development of the function for determining minimum feasible production cycle durations (introduced as Equation 1 in Chapter 3 of this proposal). However, Mabert and Pinto focus only on automated systems, and their model does not incorporate any conventional component.

Various other studies have addressed other aspects of FMS planning and scheduling, including the dynamic routing of parts in an FMS (Maimon and Choong [50]), the integration of machining and assembly scheduling in FMS (Kusiak [51]), stochastic scheduling in FMS (Gershwin [52]), flexibility in scheduling when changes occur in FMS (Stecke and Kim [53]), and hierarchial planning and control in FMS (Shin and Wilhelm [54]). However, none of these works bear directly on this study.

Lot Sizing and Scheduling in Conventional Systems

In the preceding section various studies relating to the scheduling, planning, and control aspects of FMS were reviewed. In this section, major studies relating to the same issues (i.e. lot sizing and scheduling) in a conventional environment will be reviewed. It should be noted that literally hundreds of studies addressing such issues in conventional systems have been published. A complete review of these works is unnecessary here. For an excellent review of these works see Graves [55]. In this section the focus will be only on those studies which significantly relate to this research.

A large number of studies have focused on closed job shops, which typically process multiple items through multiple work centers either for further processing or for stock. In fact, investigation of such models is one of the oldest areas in inventory planning and control research. Wagner and Whitin [57] were the first to address the issue of lot sizing under a dynamic situation. Since then a sizable literature has developed focusing on various aspects of the problem, including capacity constraints (Karmarkar, Kekre, and Kekre [58]), multiple items (Karmarkar and Schrage [59]), and multiple

Chapter 2: Literature Review 21

stages (Afentakis, et. al. [60].) A major tenet of all of these studies is the attempt to show a trade-off between too much capital tied up in too much inventory versus losses in productivity caused by producing too many small batches. However, Karmarkar [61] challenges this notion and argues that these studies "fail to completely capture the nature of the batching problem." Karmarkar argues that up to a point lead times decrease as batch sizes decrease. Beyond a certain point, however, the average time an item spends in the system increases rapidly, due to losses of capacity attributable to an excessive number of setups. The author shows that this inverse relationship between lot size and manufacturing lead time (for small batch sizes) is a very important factor in considering the total productivity of the system. The heart of this argument is conveyed in the statement that:

> Unavailability of processing capacity leads directly to deferral of work until capacity is available, causing delays in the completion of a given batch. In short, the paradigm is that manufacturing lead times are the direct result of capacity limitations. The effect of lot sizing on capacity utilization will also be apparent as an effect on lead times. [p. 410]

A major point in this study by Karmarkar [61] and other research done by the same author (see below) is that setup cost should be regarded only as loss in production time. This concept is in line with the conclusion made by Kaplan [62] that labor cost, the other portion of setup cost, should be considered "sunk cost."

In a major study, Karmarkar, Kekre, Kekre [63] develop a model for determining lot sizes in a closed job shop, taking into consideration the relationship between manufacturing lead time and lot sizes. In this study, the authors indicate that the true cost of setups (which is the main distinctive feature between a conventional system and a system with an FMI) is the inventory holding cost of parts waiting in queues which are delayed by the setups. This observation makes it clear that the cost of holding WIP inventory is a proper performance measure when comparing the conventional and the mixed conventional/FMI systems. The study by Karmarkar, Kekre, and Kekre, however, does not consider the feasibility of the lot sizes generated by their model with respect to a given production cycle. This limitation makes their model unusable for scheduling purposes and the authors, recognizing this fact, state that "Since the model takes an average or static view of the queuing phenomena in the shop, it is inappropriate for detailed scheduling decisions." (p. 295). This aspect of the problem is one of the factors considered in this study. In this study feasible production cycles that produce the required output of each item will be developed. These production cycles will be of optimum duration (assuming that direct labor involved in setups is a sunk cost) and will be based on near-optimum item sequences.

While many models of lot sizing assume demand to be deterministic, some models, such as the model developed by Askin [64], address the same issues in a stochastic demand environment. These studies are not relevant to this research in which demand will be treated as deterministic and constant.

The functional relationship between production lot sizes, the manufacturing cycle time, and the average in-process inventory is very important when the behavior of a closed shop is investigated. These issues are addressed in a study by Szendrovits [65]. Szendrovits has developed a model which treats the manufacturing lead time as a function of lot size. Using this functional relationship to determine the size of the work-in-process inventory, the author then developed an economic production quantity (EPQ). However, the model developed by Szendrovits is different from this study in at least two major respects: (1) In this study, it is the production cycle duration which determines the lot size (as indicated in Chapter III of the proposal), while in Szendrovits' study manufacturing lead time is determined by a constant lot size and (2) the major objective of the Szendrovits study is to develop an EPQ model and compare it with the traditional EOQ (or what the author calls ELQ) model, while this is not the objective of this study. The similarity of this research and Szendrovits' study is that both assume multiple setup times. Furthermore, this research was guided by the Szendrovits study in that the relationships among manufacturing cycle time, in-process inventory, and lot sizes were clarified by the latter study, although this research pursues a different set of objectives. The Szendrovits study is important in that it addresses the system relationships and behavior that is of interest in this research. However, the model focuses on conventional systems only and does not investigate the installation of a flexible work center in such a system. Several other researchers have developed economic production quantity (EPQ) models. Taha and Skeith [66] developed a model for a single-product, multi-stage conventional production system. Tersine [67] removes the limitation of single-product output, but the author develops a single-stage, multi-item (deterministic) model. The same limitation is present in a study by Graves [68] in which a system with single-stage, multiple product (stochastic) characteristics is addressed. The common point among all of these studies and this research is that the main performance criterion of the system is total inventory cost for the system. However, none of these studies address a multi-stage, multi-item closed hybrid-flow system in which an FMI is installed.

What can be concluded from the review of lot sizing and scheduling studies discussed in the preceding paragraphs is that the older studies did not address the issue of new technology such as FMI, simply because the new technology did not exist when the research was conducted. The more recent studies, on the other hand, focus on systems which are assumed to be a complete FMS or CIMS and therefore do not address the issue of introducing a flexible island into an existing conventional system. However, both branches of research provide bases for this research, and the specific contributions such studies have made with respect to the methodology of this research will be discussed in Chapter III.

Item Sequencing Considerations in Scheduling

It is clear from the foregoing discussion of literature on lot-sizing and production scheduling that a common problem in the flow shop environment is to simultaneously determine lot-sizes and product (item) sequence in a production schedule. In this section, a review of major research on the sequencing problem will be presented. As will be shown in Chapter III, the problem of determining optimum item sequences through production systems is directly relevant to this research.

Literature on sequencing of items in job shop and/or flow shop environments can be broadly classified into:

- ☞ Studies which deal with multiple-products produced on a single machine (or facility).
- ☞ Studies focusing on multiple-product, multiple-machine situations, often referred to as the n-job, m-machine environment.
- ☞ Studies dealing with the sequencing problems in an FMS environment.

Literature on sequencing in the assembly line operation environment, where the main problem is line-balancing, has also been reviewed. Among these studies, the research focusing on mixed-model assembly lines is of interest here. Although, the assembly line environment is not similar to the flow shop environment, the literature has been reviewed to determine whether concepts or methods have been developed which would be helpful in this research.

Multiple-Product, Single-Machine Studies: An early study by Delporte and Thomas [76] is typical of the research in this area. The authors developed a heuristic for the simultaneous determination of lot-sizes and sequencing of N items in a single-machine environment. The heuristic assumes deterministic demand and allows more than one batch of each item to be produced during each production cycle. The objective of the heuristic is to minimize the combined setup and inventory holding cost. In this study, items flow through a system of several machines, only one batch of each item is produced per production cycle, and the performance criterion is inventory holding cost only. Some studies have focused on more complex settings such as the case of stochastic demand (see Leachman [77]), while others have been devoted to the evaluation of alternative sequencing heuristics (see, for example, 78 and 79). Unfortunately, the heuristics developed for the multiple-product, single-machine environment cannot be applied in this study.

Multiple-Product, Multiple-Machine Studies: These studies focus on the problem of sequencing several products through several machines. All products flow through all machines in the same order. The most common objective in these studies is to minimize the makespan or flow time for a given set of production requirements. Nawaz, Enscore, and Ham [80] developed a heuristic for determining a near-optimum sequence in this environment. Vachajitpan [81] developed an integer programming formulation for determining the optimum item sequence using a branch and bound methodology. The primary differences between this line of research and this study, are (1) all products flow through all machines in the same sequence, and (2) the performance criterion is the makespan. Due to these differences, the methodologies developed in these studies are not usable in this research.

Another line of research in the multiple-product, multiple-machine environment focuses on sequence-dependent setup times or costs. In the case of sequence-dependent setup times (see Smith-Daniel and Ritzman [82]), the objective is to minimize past due penalties or weighted completion times. In the case of sequence-dependent setup costs (see Galvin [83]), the objective is to minimize the sum of inventory holding costs plus setup costs. Due to the performance criteria and assumptions concerning system structure, these studies are not helpful with respect to this research.

Sequencing Studies in FMS Environment: There are not many studies focusing on sequencing in the FMS environment. However, some of the sequencing heuristics and algorithms developed for conventional systems have been evaluated in the FMS environment. In a recent study, Co, Jaw, and Chen [84] compared five sequencing rules, one of which was developed by the authors themselves. The purpose of the study was to determine which of the rules performed better with respect to minimizing the flow time of production. The objective of minimizing inventory holding cost has not been addressed with respect to item sequencing in the FMS environment.

Sequencing in assembly-line environment: Many studies have been devoted to the problem of determining the optimal launching scheme for products on assembly lines that produce more than one product. Wester and Kilbridge [85], Dar-El and Cucuy [86], and Church [87] have all studied the mixed-model assembly line balancing problem. Thomopolous [85] considered two cases of assembly line balancing models: (1) batch assembly and (2) mixed-model assembly. In the first case, the sequencing of items (products) is not an issue. In the mixed-model case, the sequencing of products is addressed and two rules are investigated: (1) variable-rate launching and (2) fixed-rate launching. However, the objective is balancing the workload among the work stations rather than reducing inventory holding cost. Furthermore, all products follow the same sequence of work stations, which is clearly not the case in this research. Similarly, Em Dar-El and Cucuy [86] and Church [87] address the assembly line balancing problem. However, as in Thomopolous' study, the main objective of these studies is to increase efficiency of the production line by balancing the workload among the work stations (i.e., minimize the idle time). Unfortunately, the rules developed in these studies are not applicable to this research.

In summary, although item sequencing has been an issue in previous scheduling research, the assumptions concerning the structure of the system studied and the objectives of the models developed do not match the assumptions and objectives of this research. It is noteworthy, however, that most of the methodologies developed in previous research for dealing with sequencing problems involve heuristics rather than optimizing techniques. Thus, in Chapter III, a heuristic is presented for determining near-optimum item sequences for the system under study.

In Brief

In this chapter, the literature relating directly to this study was reviewed. In the first section, the literature on the benefits of flexible manufacturing systems was discussed. Most of the studies in this vein are anecdotal. The second section reviewed the growing body of literature on problems involved in implementing new manufacturing technologies, such as FMS. The effective integration of flexible and conventional subsystems was shown to be a major problem. Analytical and mathematical models dealing with the design and control of FMS were reviewed next, followed by a review of studies focusing on lot sizing and scheduling issues in such systems. In the next section of this chapter the literature on lot sizing and scheduling issues in conventional systems was discussed. Finally, previous research on item sequencing issues in scheduling was reviewed. In all cases, the relationship between a particular work and this study was indicated. The conclusion was that no study has approached the specific problem that this research will address, which is the impact of introducing an FMI into an otherwise conventional manufacturing system. Several authors, however, have emphasized the importance and complexity of this problem.

CHAPTER III

RESEARCH DESIGN AND METHODOLOGY

The Choice of Methodology

The choice of methodology for a research project is primarily dependent on: (1) the objectives of the research, and (2) the complexity of the situation being investigated. Sarin and Wilhelm [70], in their comparison of various research tools to study FMS, indicate that simulation is suitable for many of the operating and planning problems posed by such systems. These problems include: factory layout design, part routing, machine tooling, dispatching rules, evaluation of machine breakdowns, and scheduling. Concurring with this point, Jain and Foley [71] state that "The complex interactions in an FMS make it nearly impossible to closely model it analytically. Simulation offers a powerful approach for FMS modelling and is being extensively used." Similarly, Shannon and Phillips [74], while discussing the benefits of simulation as a research tool, indicate that this tool is a very powerful one to be used in the design and operation of complex systems. An observation which bears directly on the choice of methodology in the present study is made by Wortman [75], who indicates that "For any given potential FMS application, the number of feasible FMS configurations is enormous. However, identifying the 'best' configuration is virtually impossible without a method for evaluating and comparing alternatives."

Although, the purpose of the present study is not designing or finding the "best" configuration of an FMS, the nature of the study involves comparison of several possible configurations composed of conventional and flexible work centers. The nature of the simulation to be employed is somewhat unique. Although system behavior (i.e. state variables) will be projected through time, the projections will be developed analytically, rather than by having a model step through time. In terms of

Chapter 3: Research Design & Methodology 29

typical classifications of simulation models, the simulation involved in this study will be deterministic and continuous.

Definitions

Before continuing with the details of the methodology and the procedures to be used in the study, it is necessary to define the concepts and terminology used in the research. It should be noted that in many areas of factory automation, there is no universal set of definitions on which even the majority of researchers or users/suppliers agree.

Flexible Manufacturing Systems (FMS)

Although there are many definitions for FMS, in this research the definition given by Ranky [75] will be used. He defines FMS as:

> *a system dealing with high level distributed data processing and automated material flow using computer controlled machines, assembly cells, industrial robots, inspection machines, and so on, together with computer integrated material handling and storage systems.* (p. 1)

Buzacott and Yao [76] provide a conceptual design for FMS which is depicted in Figure 2. This conceptual design illustrates the components of FMS as described in Ranky's definition.

Flexible Manufacturing Island (FMI)

As used in this study, FMI refers to a subsystem (work center) within a conventional, non-automated system that has been converted into a flexible work center via computer-controlled automation. The distinguishing characteristic of an FMI is that it is capable of switching from one operation to another with negligible setup time. This characteristic makes it possible to have unit flow among work centers, as opposed to batch flow, which would be the mode in

Flexible Manufacturing Islands 30

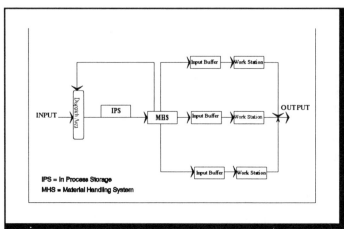

Figure 2. A Conceptual Design for a Flexible Manufacturing System

a totally conventional system. Although not of concern in this study, it should be noted that a flexible work center is capable of performing only a limited number of different operations.

Variables and Experimental Design

The dependent and independent variables of this study are defined as follows:

Dependent Variable

Total inventory holding cost per year for the entire system is the dependent variable. Setup cost is not considered a relevant dependent variable in this research. As indicated by Kaplan [62], the main element of setup cost is setup time, and Karmarkar [61] points out that setup time should be considered as lost production time rather than as an incremental cost. Setup times for individual operations will be treated as system parameters in the simulations.

Chapter 3: Research Design & Methodology   *31*

Independent Variables

The primary independent variable in this study is the presence (versus absence) of one FMI in the otherwise conventional system along with the position of the FMI in the system.

The effect of introducing the FMI in various positions will be evaluated separately for three different system configurations. The results will be evaluated for sensitivity to three intervening variables, which are:

1. System Capacity Utilization
2. Setup Time to Run Time Ratio: This variable refers to the ratio of setup time required for a batch of a given item at a given work center to the processing (run) time per unit at that work center.
3. Rate of Increase in Holding Cost: This variable refers to the rate of increase in holding costs as items progress through successive work centers.

Two levels of each of the above intervening variables (i.e., low and high) will be compared to a base (medium) level. The method for establishing these levels will be explained later in the chapter. Table 3 shows a matrix for recording the results (total inventory holding cost/year) for each combination of the independent variables. In Table 3, the "base" column refers to the situation in which each of the three intervening variables are set at their medium levels. As shown in Table 2, only one non-common work center for each path operation number in each system configuration will be converted into FMI.

The criterion for selecting the non-common work center for conversion into FMI is the production cycle duration. The method for selecting the non-common work center for conversion to FMI will be explained later in this chapter.

Table 2.. Work Center Conversion to FMI

Configuration	Path Operation #	Work Center to Convert
I	2	C1 or C2 or C3 or C4
II	1	A1 or A2
II	3	C1 or C2
III	1	A1 or A2 or A3 or A4

Table 3
Results Matrix

Config.	FMI Position	Base	SENSITIVITY ANALYSIS					
			Parameters to be tested					
			Capacity Utilization		Setup/Run Time Ratio		Rate of Increase in Holding Cost*	
			Low	High	Low	High	Low	High
I	No FMI							
	WC B							
	WC Ci							
II	No FMI							
	WC B							
	WC Ci							
III	No FMI							
	WC B							
	WC Ai							

*Column heading refers to holding cost parameters (dollar per unit per day)

Assumptions and Description of the Model

The system modeled in this study is characterized by the following assumptions and descriptions:

1. Three system configurations are considered, as shown earlier in Figure 1.
2. The system is completely deterministic.
3. The system produces eight different items.
4. Demand for each item is constant.
5. System capacity is adequate to meet demand for all items, and no backlogging of demand is allowed, assuming a production cycle duration of adequate length (i.e., production batches of adequate size) is established.
6. Paths of all items are predetermined. Thus, the path of items is a technological constraint. All items flow in the same general direction through the system. More precisely, all flows between any two work centers are in the same direction. Four paths are possible for each system configuration and two items follow each of the four paths. Any given work center corresponds to the same operation number in each path in which that work center appears. Each system configuration involves a "common" work center that appears in the paths of all eight items.
7. Only <u>batch flows</u> between conventional (i.e., non-FMI) work centers are allowed. In other words, items will not flow from one conventional work center to another until the whole batch is completed at the first work center. <u>Unit flow</u> is allowed, however, to and from an FMI. This means that as individual items are processed at a conventional work center, they will immediately flow to the next work center on their path, if that work center is FMI. Similarly, items processed at an FMI can flow unit-by-unit to the next work center in their path.
8. Setup times are independent of item sequences.
9. Holding cost rates increase as items are processed through successive operations.

Flexible Manufacturing Islands   34

10. No breakdowns of equipment or quality problems are allowed.
11. Results for all system configurations and FMI positions will be evaluated under the optimum production cycle duration and a near-optimum item sequence (i.e., the sequence of the items as they pass through the common work center) for that system.
12. System parameters will be selected such that for the totally conventional system (i.e., no FMI) the common work center will be the determinant of system capacity. Therefore, the common work center will determine the minimum feasible cycle duration for the system which will be the optimum cycle duration with respect to inventory holding cost. The relationships between system parameters, system capacity, and optimum cycle duration are explained later in this chapter.
13. The system produces one and only one batch of each item during each production cycle duration.

Detailed Procedure of the Study

A simplified system is used to illustrate the methodology of this research. This system is composed of three work centers with the configuration depicted below:

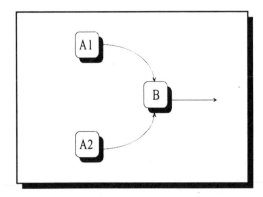

Chapter 3: Research Design & Methodology 35

The system produces four items, numbered 1 through 4. Table 4 shows the annual demand (D) for the items and indicates the paths for each item. For each item the table gives:

1. The setup time (s) in days/batch, assuming conventional work centers.
2. The production run time (t) in days/unit. The production rate (p) can be computed as $1/t$.
3. The holding cost rate (h) in dollars/unit/day for items immediately following the operation.

Table 4
Parameters of the Simplified System

Production Information		Operation No. 1						Operation No. 2		
		Work Center A1			Work Center A2			Work Center B		
Item	Demand	s	t	h	s	t	h	s	t	h
1	300	4	.25	.06	....	....	....	3	.10	.12
2	480	....	....	....	5	.18	.05	2	.20	.10
3	780	3	.20	.08	....	....	....	4	.08	.15
4	600	....	....	....	2	.22	.10	1.5	.14	.14

Conventional System

Base Case

The first step is to calculate the minimum feasible production cycle duration, C_{min}, for each work center. Equation #1 which is adapted from Mabert and Pinto [56] is used to determine this value.

$$C_{min} = \frac{\sum_{i=1}^{n} s_i}{1 - \sum_{i=1}^{n}\left(\frac{d_i}{p}\right)} \quad (1)$$

There d_i represents the demand rate expressed in units per day; the summations of s_i and d_i/p_i include all items that are processed at work center j, and $d_1 + d_2 + \ldots d_i < p_1 + p_2 + \ldots p_i$. The largest C_{min} value is the minimum feasible production cycle duration for the conventional system. Given the assumption of only and only one batch production of each item in a given production cycle duration, it is clear that the minimum feasible production cycle duration is the optimum cycle duration with respect to the minimization of inventory holding cost. In this example (assuming 360 days per year) the production cycle durations for work centers A1, A2, and B are 19.53, 17.80, and 30 days respectively. Thus, the minimum feasible and therefore optimum production cycle duration for the system, C_{opt}, is 30 days where C_{opt} is defined as

$$C_{opt} = Max(C_{min\,j}) \quad (2)$$

Based on C_{opt}, the batch sizes (Q_i) required to meet the annual demand for each item will be determined using the following function:

Using the annual demands as given above for items 1, 2, 3, and 4, the batch sizes for these items are 25, 40, 65, and 50 units

Chapter 3: Research Design & Methodology   37

$$Q_i = C_{opt} \, d_i \quad (3)$$

per batch respectively.

The next step is to specify the sequence of items as they are processed at the common work center (i.e., work center B). As previously explained, the sequence can affect the inventory holding cost for the system. In order to compare system performance with respect to inventory holding cost before and after the introduction of an FMI, the optimum (or near-optimum) item sequence should be used. For a system of the type involved in this research that produces N items (N=4 items in the example), the number of possible item sequences would seem to be N! (4! = 24 sequences in this example). However, when viewed on a cyclical basis, it becomes apparent that the number of unique sequences is actually only N!/N or (N-1)!. For example, the sequences 1-2-3-4, 2-3-4-1, 3-4-1-2, and 4-1-2-3, all result from the cyclical pattern shown below and they account for only one unique sequence. Thus, the number of unique sequences for this example is six (or 3!), and they can be identified as follows:

1. 1-2-3-4   4. 1-2-4-3
2. 1-3-2-4   5. 1-3-4-2
3. 1-4-2-3   6. 1-4-3-2

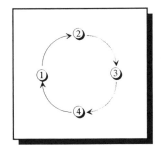

For the purposes of this example, the procedure for projecting inventory behavior and evaluating holding cost for only one of these sequences, which is the sequence 1-2-3-4 will be demonstrated. However, the total annual inventory holding cost that results from each of six unique sequences will also be shown. Upon examining those results, the discussion of selecting optimum (or near-optimum) item sequences will be resumed.

To illustrate the methodology for projecting inventory behavior and evaluating holding cost for the example, the item sequence 1-2-3-4 will be used. Beginning with the common work center B, the production cycle starts by setting up the work center to process item #1. It takes three days to set up the work center. Then the processing of item #1 begins, and it takes 2.5 days to process the batch size of 25 units at the processing rate of 10 units per day (refer to Table 4 on page 35 for the parameters of this example). When the processing of item #1 is finished on day 5.5, work center B will be set up to process item #2. This setup time is 2 days and, therefore, processing of item #2 begins on day 7.5 and continues at the rate of 8.33 units per day until the batch of 40 units is completed on day 12.3. In the same manner items #3 and #4 will be processed. The processing times and inventory behavior at work center B for all items of the simplified example are shown in Figure 3. Figure 4 is the Gantt chart showing the scheduling of work centers B, A1, and A2 for item sequence 1-2-3-4.

To explain the inventory behavior in Figure 3, item #1 will be used as an example. Processing of item #1 begins at work center B on day 3 (after setup is completed) at which time the inventory of item #1 has just run out. Item #1 is processed at a rate of 10 units per day. However, as items are being processed, they are also being used up (demanded) at a constant rate of .833 units per day (300 units per year/360 days). The difference between the processing rate and consumption rate of item #1 leads to the buildup of item #1 inventory following work center B at a rate of 9.17 units per day for 2.5 days (time required to produce the batch of 25 units). This results in a peak inventory of 22.9 units of item #1 following work center B. After processing of item #1 ceases at work center B, the inventory of item #1 continues to be used up at the rate of .833 units per day.

Chapter 3: Research Design & Methodology 39

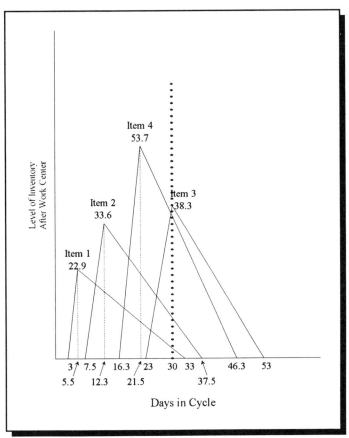

Figure 3. Inventory Behavior Following Work Center B

Flexible Manufacturing Islands 40

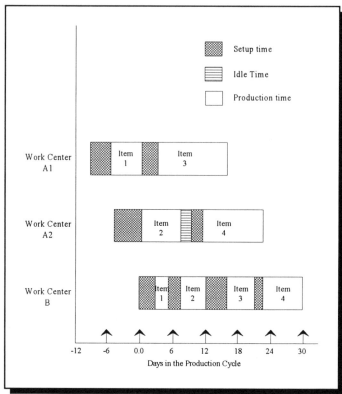

Figure 4. Gantt Chart for Scheduling of Work Centers B, A1, and A2

Chapter 3: Research Design & Methodology   41

The inventory of item #1 reaches zero three days into the next cycle (day 33) when processing of item #1 is due to begin again. Similar behavior is observed for items 2, 3, and 4.

The inventory levels and processing times for work center A1 and A2 are shown in Figures 5 and 6. Although the inventory and processing behavior at non-common work centers (i.e., A1 and A2)follow similar patterns, they are somewhatdifferent from the common work center B. Assuming conventional work centers, items are not used by work center B as soon as they are processed by A1 or A2. Only after an entire batch of an item has been processed at work center A1 or A2, does the batch flow to work center B. For example, processing of item #3 starts on day 3.3 at work center A1 and continues until the whole batch of 65 units is completed. Note that the batch for item #3 is completed at work center A1 exactly when processing of that item is to begin at work center B (day 16.3). Such timing minimizes the cost of holding work-in-process inventory of item #3 following work center A1. Due to the sequencing of items, however, there is a conflict of processing schedules between items #1 and #3 at work center A1. In other words, the process should ideally start such that the batch of 25 units will be completed exactly when processing of that item is to begin at work center B (day 3). However, implementing this ideal processing schedule would delay the processing of item #3 and its flow to work center B. Therefore, the solution is to process one of the items at work center A1 earlier than the ideal time. However, this implies carrying inventory following work center A1 for a longer period of time, thus increasing inventory cost. Therefore, the obvious priority of processing of items when such conflict occurs would be to process those items which have the lowest batch holding cost earlier than they are actually needed by the next operation. For example, item #1 has a unit inventory holding cost of $.10/unit/day and the batch size of 25 units. Therefore, the batch holding cost is $2.50/batch/day. The same parameters for item #3 are $.12 /unit/day, 65 units, and $7.80/batch/day respectively. Thus, it is more economical to process item #1 early at work center A1 and carry it until it is needed by work center B. This explains the flat section of the inventory graph for item #1 following work center A1 (see Figure 5). No such scheduling conflict occurs at work center A2. Items #2 and #4 are completed at work center A2 exactly when processing of these items begins at work center B (day 7.5 and 22.5 respectively).

Once the inventory behavior of all items at all work centers has been determined, the inventory holding cost for the system during a complete production cycle is determined simply by calculating the areas under the inventory behavior graphs and multiplying the amount of inventory unit-days by the inventory holding cost of each item per day. The inventory of item #1 in work center A1 is 176.88 unit-days per 30-day cycle. Since the holding cost for item #1 following work center A1 is $.06 per unit per day, the inventory holding cost of item #1 following work center A1 is $10.61 per cycle. Calculated similarly, inventory for item #3 following this work center is 591.50 unit-days, the holding cost is $.08 per unit per day, and the total holding cost is $47.32. The total inventory cost for work center A1, therefore, is $57.93 for the cycle duration of 30 days. The annual inventory holding cost for items #1 and #3 after work center A1 (again assuming 360 days/year) is $695.16. Calculating the inventory holding cost for items #2 and #4 following work center A2 and for all four items following work center B is done in the same manner. The results are shown below in Table 5. The total annual inventory holding cost for the entire system, given the item sequence of 1-2-3-4, is $4893.72. Table 5 also shows the results for the other five unique item sequences that were identified previously.

Chapter 3: Research Design & Methodology 43

Table 5
Annualized Total Inventory Holding Cost For Various Item Sequences

Sequence	WC A1	WC A2	WC B	Total
1-2-3-4	$695.16	$684.00	$3514.56	$4893.72
1-3-4-2	817.56	684.00	3514.56	5016.12
1-3-2-4	817.56	938.40	3514.56	5270.52
1-2-4-3	646.56	844.80	3514.56	5005.92
1-4-3-2	673.56	684.00	3514.56	4872.12

Note that the total inventory holding costs are significantly less for the sequences 1-2-3-4 and 1-4-3-2 than for the other four sequences. These two sequences fall into a special class referred to as "alternating sequences" in this research. An alternating sequence is defined as follows:

Given that R different paths are possible within a system, an item sequence will qualify as an alternating sequence if for every set of R consecutive items in the sequence, no two items in the sequence follow the same path.

Our simplified example involves two paths, which are A1-->B and A2-->B. For the item sequences 1-2-3-4 and 1-4-3-2, any two consecutive items within the sequences follow different paths. This is not true for the other four sequences. For example, in the sequence 1-3-2-4, items #1 and #3 (which are consecutive in the sequence) both follow the path A1-->B. Items #2 and #4 also follow the same path (i.e., A2-->B).

As illustrated by the simplified example, alternating sequences tend to yield lower inventory holding costs than non-alternating sequences. The reason for this cost advantage is that alternating sequences minimize interference at the non-common work centers (such as the interference that occurred at work center A1 in our example, which required

Flexible Manufacturing Islands 44

that item #1 be held in inventory for a brief period of time before processing of that item could begin at work center B). Thus, the optimum item sequence with respect to inventory holding cost for each system configuration in this research is expected to be an alternating sequence.

For each of the systems to be addressed in this study, the number of items (N) is eight, and the number of paths (R) is four. Thus, the number of unique sequences can be calculated as:. $(N-1)! = 7! = 5,040$. Of these unique sequences, 48 will qualify as alternating sequences, and they are listed in Table 6. This set of alternating sequences is based on the fact that in each case, items #1 and #5 will follow the same path as will items #2 and #6, items #3 and #7, and items #4 and #8.

As indicated in Chapter II, although considerable research has been conducted on various sequencing problems, no methodology for determining optimum sequences has been developed which is directly relevant to or helpful in this study. For the purposes of this study, five of the 48 unique alternating sequences will be selected randomly. These five sequences will be evaluated for all cases in all system configurations. In each case, the sequence resulting in the lowest total inventory holding cost will be considered the near-optimum sequence and will determine the holding cost for that case. In many cases, as indicated in Chapter IV, a clearly optimum item sequence is found among the five randomly selected item sequences. Such clearly optimum item sequence produces no production schedule conflict at any work center in the system.

Chapter 3: Research Design & Methodology   45

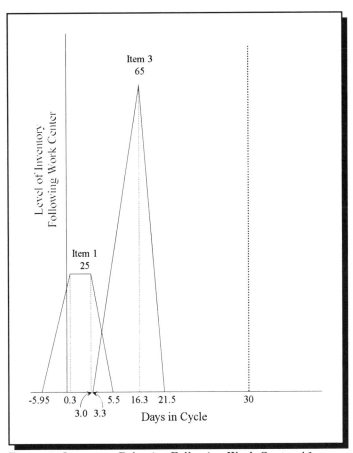

Figure 5. Inventory Behavior Following Work Center A1

Flexible Manufacturing Islands 46

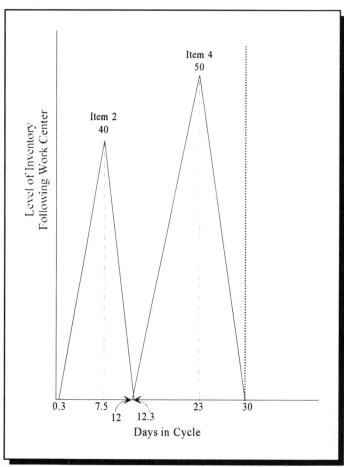

Figure 6. Inventory Behavior Following Work Center A2

Chapter 3: Research Design & Methodology 47

Table 6
List of Unique Alternating Sequences in this Study

No.	Item Sequence	No.	Item Sequence	No	Item Sequence
1	1-2-3-4-5-6-7-8	17	1-7-4-2-5-3-8-6	33	1-8-3-6-5-4-7-2
2	1-3-2-4-5-7-6-8	18	1-4-2-7-5-8-6-3	34	1-6-8-3-5-2-4-7
3	1-4-3-2-5-8-7-6	19	1-2-3-8-5-6-7-4	35	1-3-8-6-5-7-4-2
4	1-2-4-3-5-6-7-8	20	1-3-2-8-5-7-6-4	36	1-8-6-3-5-4-2-7
5	1-3-4-2-5-7-8-6	21	1-8-3-2-5-4-7-6	37	1-2-7-8-5-6-3-4
6	1-4-2-3-5-8-6-7	22	1-2-8-3-5-6-4-7	38	1-7-2-8-5-3-6-4
7	1-6-3-4-5-2-7-8	23	1-3-8-2-5-7-4-6	39	1-8-7-2-5-4-3-6
8	1-3-6-4-5-7-2-8	24	1-8-2-3-5-4-6-7	40	1-2-8-7-5-6-4-3
9	1-4-3-6-5-8-7-2	25	1-6-7-4-5-2-3-8	41	1-7-8-2-5-3-4-6
10	1-6-4-3-5-2-7-8	26	1-7-6-4-5-3-2-8	42	1-8-2-7-5-4-6-3
11	1-3-4-6-5-7-8-2	27	1-4-7-6-5-8-3-2	43	1-6-7-8-5-2-3-4
12	1-4-6-3-5-8-2-7	28	1-6-4-7-5-2-8-3	44	1-7-6-8-5-3-2-4
13	1-2-7-4-5-6-3-8	29	1-7-4-6-5-3-8-2	45	1-8-7-6-5-4-3-2
14	1-7-2-4-5-3-6-8	30	1-4-6-7-5-8-2-3	46	1-6-8-7-5-2-4-3
15	1-4-7-2-5-8-3-6	31	1-6-3-8-5-2-7-4	47	1-7-8-6-5-3-4-2
16	1-2-4-7-5-6-8-3	32	1-3-6-8-5-7-2-4	48	1-8-6-7-5-4-2-3

Sensitivity Analysis

In this phase of this research, three variables will be manipulated independently in order to determine their impact on the inventory holding cost of the system. The optimal (or near-optimal) sequence of items as determined in the previous phase will be used. The optimum cycle duration will change, however, as demand rates and setup times are altered. The three variables, as defined previously, are:

1. System capacity utilization
2. Setup time to run time ratio
3. Rate of increase in holding cost

System Capacity Utilization

System capacity utilization will be calculated at the work center for which the minimum feasible cycle duration (C_{min}) as determined by Equation 1 is largest and therefore determines the optimum cycle duration (C_{opt}) for the system. The measure of system capacity utilization (U) will be the ratio:

$$U = \sum_{i=1}^{n}\left(\frac{d_i}{p_i}\right) \quad (4)$$

where the summation is performed over all items that pass through that work center. In the base case for the simplified example, work center B (the common work center) determines the cycle duration for the system. Using the base case parameters for work center B shown in Table 4, the utilization ratio is .65. Clearly, this measure of utilization does not include time devoted to setups. Note that if the ratio reached the value of 1.0, the denominator of Equation 1 would become zero, and no feasible cycle duration would exist. In other words, work center B would not be able to keep pace with demand if processing were interrupted for setups.

As indicated earlier, one of the limiting assumptions of this study is that for each system configuration the system parameters will be selected such that for the conventional system, the common work center will be the determinant of system capacity and therefore of the optimum production cycle duration for the system. Furthermore, the base case parameters will be set such that capacity utilization will be .65, as it was in the simplified example.

To obtain the high level of capacity utilization, the demand for each item will be increased by 30 percent, which will increase the overall capacity utilization (U) by 30 percent to approximately .85. Similarly, the low level of capacity utilization will be obtained by decreasing the demand for each item by 30 percent, resulting in an overall capacity utilization (U) of about .45.

Table 7 shows the annual demand level for the four items in the simplified example at the base (medium) level, low level, and high level of system capacity utilization rates.

Chapter 3: Research Design & Methodology

Setup Time to Run Time Ratio

The setup times for the items shown in Table 4 represent the base (medium) ratio of setup times to run times for the simplified example. The base level setup times will be increased and decreased by 30 percent to achieve the high and low ratio of setup times to run times for all items at all work centers. Table 8 shows the setup times for the four items at work centers A1, A2, and B in the simplified example which would produce the base (medium), low, and high ratios of setup times to run times.

Table 7
Annual Demand Rates for Sensitivity Analysis

	ANNUAL DEMAND LEVEL		
Item	Low	Medium	High
1	210	300	390
2	336	480	624
3	546	780	1014
4	420	600	780

Table 8
Setup Times for Sensitivity Analysis

	SETUP TIMES								
	WC - A1			WC - A2			WC - B		
Item	Low	Med.	High	Low	Med.	High	Low	Med.	High
1	2.80	4.00	5.20				2.10	3.00	3.90
2				3.50	5.00	6.50	1.40	2.00	2.60
3	2.10	3.10	3.90				2.80	4.00	5.20
4				1.40	2.00	2.60	1.05	1.50	1.95

Flexible Manufacturing Islands   50

Rate of Increase in Inventory Holding Cost

This variable will be manipulated such that for each item, the mean inventory holding cost (averaged across all operations in the path of the item) remains constant. The parameters in Table 4 provide the base level for the simplified example. To create the low rate of change, the holding cost rate for the <u>first</u> operation in the path will be <u>increased</u> by 15 percent, and holding cost rate for the <u>last</u> operation will be <u>decreased</u> by an equal dollar amount. To obtain the high rate of change, the holding cost rate for the <u>first</u> operation will be <u>decreased</u> by 15 percent and holding cost rate for the <u>last</u> operation will be <u>increased</u> by an equal dollar amount. Table 9 shows holding cost rates of change in the simplified example. Note that the average holding cost/unit for all three cases (i.e., low, medium, and high rete of increase) remain constant; that is .09, .075, .115, and .120 for items 1, 2, 3, and 4 respectively.

Table 9
Inventory Holding Cost for Sensitivity Analysis

Item	Low Rate of Increase			Medium Rate of Increase			High Rate of Increase		
	WCA 1	WCA 2	WCB	WCA 1	WCA 2	WCB	WCA 1	WCA 2	WCB
1	.069	....	.111	.060	....	.120	.051	....	.129
2	....	.058	.092	....	.050	.100	....	.042	.108
3	.092	....	.138	.080	....	.150	.068	....	.162
4	....	.115	.125	....	.100	.140	....	.085	.155

Common Work Center Converted to FMI

Base Case

This section of the investigation begins by converting the common work center (work center B in the simplified example) into an FMI. As discussed previously, the primary effect of this conversion is that the setup times for the items processed by the common work center will become negligible. Another

Chapter 3: Research Design & Methodology   51

effect is the possibility of unit flow from work centers A1 and A2 to work center B.

Since the minimum feasible cycle duration (C_{min}) for conventional work center B had determined the optimum cycle duration (C_{opt} = 30 days) for the conventional system, the optimum cycle duration changes when work center B is converted into an FMI. Once one work center has been converted into an FMI, that work center is no longer a factor in determining the optimum cycle duration for the system (since it has zero setup times). In fact, in the simplified example, once the common work center (B) has been converted into an FMI, the non-common work centers (A1 and A2) can operate on two different production cycles with minimum feasible (optimum) cycle duration of 19.53 and 17.8 days respectively.

For the purposes of this example, the inventory holding costs will be determined using the same item sequence that was used in the previous phase of the investigation (i.e., 1-2-3-4). The method of calculating the inventory holding costs is the same with the mixed conventional-FMI systems as the method used for conventional systems. However, due to the fact that work center B no longer determines the optimum production cycle for the system, and that work centers A1 and A2 can now operate on independent production cycles, the inventory behavior following work centers A1 and A2 is somewhat different. Figure 7 and Figure 8 show the inventory behavior following these work centers respectively. The basis for inventory build-up following these work centers is similar to the conventional system case. Figure 7 will be used to illustrate the behavior of inventory for item #1 following work center A1. Since the cycle duration for work center A1 is now 19.53 days, the batch sizes for item #1 and #3 are reduced to 16.3 and 42.3 respectively (we will allow fractions of units for ease and consistency of calculation). After setting up the work center to process item #1 (4 days), the processing of the item begins at a the rate of 4 units/day. However, since work center B is now a flexible work center and can produce individual items to exactly meet demand, item #1 will also be used up following work center A1 at the rate of .833 units/day (which

Flexible Manufacturing Islands 52

is equal to demand rate for the item). The difference between the processing rate of 4 units/day and consumption rate of .833 units/day results in an inventory build up which peaks at 12.89 units on day 8.07. After this point, inventory is used at a rate of .833 units/day until it reaches zero on approximately day 24, when it is time to start processing item #1 again. This explanation of inventory behavior also applies to items #2, #3, and #4. The build-up of inventory after work centers A1 and A2 when the common work center (B) has been converted into an FMI results in the annual inventory holding cost of $980.20 for the system. The breakdown of this total inventory holding cost is shown in Table 10.

Sensitivity Analysis

The next step in this research is to do sensitivity analysis with the mixed conventional/FMI system, when the common work center has been converted into an FMI. The method of manipulating the variables and calculating the resulting effect on inventory holding costs is the same as illustrated in the case of the conventional systems.

Table 10
Annual Inventory Holding Cost After Common Work Center Has Been Converted to FMI

Item	WC-A1	WC-A2	WC-B	Total
1	$139.17	...	$0.00	$139.17
2	...	$158.78	0.00	158.78
3	344.70	...	0.00	344.70
4	...	337.55	0.00	337.55
Total	$483.87	$496.33	0.00	$980.20

Chapter 3: Research Design & Methodology 53

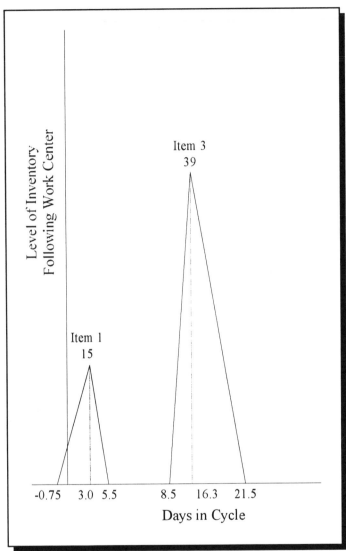

Figure 7. Inventory Behavior Following Work Center A1 When Work Center B Has Been Converted to FMI

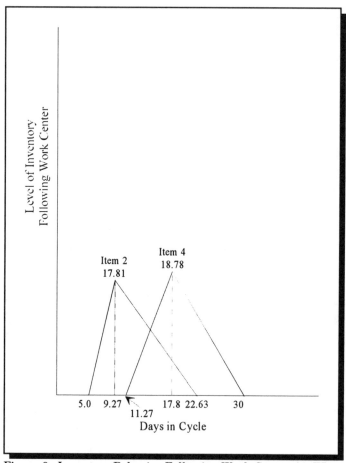

Figure 8. Inventory Behavior Following Work Center A2, When Work Center B Has Been Converted to FMI

Non-Common Work Center Converted to FMI

Base Case

The next phase is to convert one of the non-common work centers (A1 or A2 in the simplified example) into an FMI. The criterion for choosing the one non-common work center for

Chapter 3: Research Design & Methodology   55

conversion is the inventory holding cost for the work centers in the totally conventional system. As shown in Table 5, the holding costs for work centers A1 and A2 were $695.16 and $684.00 respectively. Therefore, work center A1 is selected for conversion to FMI. The logic of this selection criterion is that the work center with the highest holding cost offers the greatest potential savings.

Since the common work center is conventional, the entire system must again operate on a 30-day production cycle with the same batch sizes used in the case of totally conventional system. The inventory behavior and holding costs following work centers A2 and B ar exactly the same as in the totally conventional system (see Figures 3 and 5). Even though work center A1 is an FMI, it must still produce in batches, since work center B consumes the output of A1 in batches. However, unit flow is now possible between A1 and B. Therefore, it is not necessary to build up inventories of items #1 and #3 equal to the batch sizes for those items between work centers A1 and B. On the other hand, because the production rates at work center B are higher than the processing rates at work center A1 (for both items #1 and #3), it is necessary to build up some work-in-process inventory between work center A1 and work center B in order to prevent starvation of work center B. The size of the required buffer inventory can be determined for each item by the following function:

$$B_{A1i} = (p_{Bi} - p_{A1i})\left(\frac{Q_i}{p_{Bi}}\right) \quad (5)$$

Where: B_{A1i} = Buffer size for item i following work center A1
p_{Bi} = Processing rate for item i at work center B
p_{A1i} = Processing rate for item i at work center A1
Q_i = Batch size for item i as determined by system's cycle duration.

Thus, the buffer inventories for items #1 and #3 following work center A1 are 15 and 39 respectively. The buffer for each item must be built up before work center B starts processing that item. The inventory behavior following work center A1 is shown in Figure 9, and the resulting annual inventory holding costs for the system are shown in Table 11.

Table 11
Annual Inventory Holding Cost
After Work Center A1 Has Been Converted to FMI

Item	WC - A1	WC - A2	WC - B	Total
1	$25.65		$494.64	$520.29
2		$144.00	604.80	748.80
3	243.36		1449.96	1693.32
4		540.00	965.16	1505.16
Total	269.01	684.00	3514.56	4467.57

Chapter 3: Research Design & Methodology 57

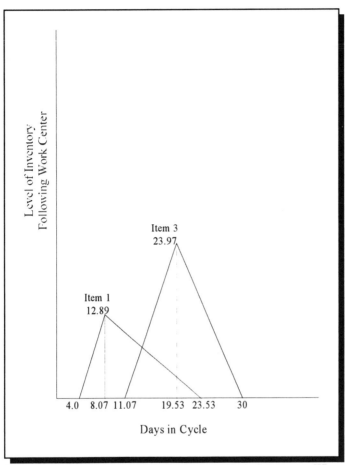

Figure 9. Inventory Behavior Following Work Center A1, When Work Center A1 Has Been Converted to FMI

Flexible Manufacturing Islands 58

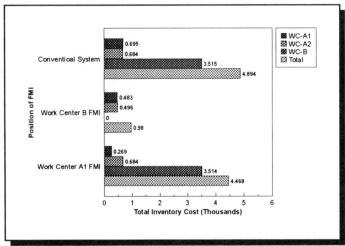

Figure 10. Comparison of Annual Inventory Holding Cost: Conventional System vs. FMI Position

Sensitivity Analysis

The final step is to perform sensitivity analysis on the mixed conventional/FMI system when work center A1 has been converted into an FMI. The method of manipulating the variables and calculating the resulting impact on inventory holding cost is the same as explained for the case of the totally conventional system.

In Brief

This chapter has explained and illustrated the methodology used in this study. The results of the simplified example (without sensitivity analyses) are summarized in Table 12. Comparison of annualized inventory holding costs when: (1) the system is totally conventional, (2) when the common work center has been converted into FMI, and (3) when one

Chapter 3: Research Design & Methodology 59

non-common work center has been converted into FMI is shown in Figure 10.

Table 12
Summary of Simplified Example Results For Item Sequence 1-2-3-4

Case	Annual Inventory Holding Cost
Conventional System	$4,893.72
Bottleneck Work Center Converted to FMI	980.20
Non-Bottleneck Work Center Converted to FMI	4,467.57

Clearly, the conversion of a single work center into an FMI can have a major impact on the inventory holding costs for an otherwise conventional system, and the relative magnitude of that impact depends, in part, on the position within the system of the converted work center. Moreover, the operating discipline of the system (in terms of production cycle durations, batch sizes, work centers' interdependence, etc.) is altered depending upon the location of the FMI. This research will provide a more complete analysis of these impacts and relationships.

CHAPTER IV

RESULTS

Introduction

The research was carried out in three phases. Each phase focused on one of the system configurations described in Chapter I. The results of the research are presented following these phases. For each phase, the structure of the system configuration is reviewed with the aid of a schematic diagram, and the base case parameters of the system are presented in tabular form. Then, the results of the investigation are presented in the following order:

Case 1: Conventional system

   A. Base case
   B. Sensitivity analysis

Case 2: System with the bottleneck work center converted into FMI

   A. Base case
   B. Sensitivity analysis

Case 3: System with a non-bottleneck work center converted into FMI.

   A. Base case
   B. Sensitivity analysis

Configuration II will involve four cases rather than three. Case 3 will examine the system with a back-end non-bottleneck work center converted into an FMI, and case 4 will address the system with a front-end non-bottleneck work center converted into an FMI. Due to the complexity of the investigation for configuration II and the fact that numerous relationships will be explained by making reference to configurations I and III, the discussion of results for configuration II are presented after the other two configurations. The discussion of each

Chapter 4: Results 61

configuration will end with a summary of results for the totally conventional system and for the various FMI positions. This discussion will be supplemented by tabular and graphical comparisons of inventory holding costs resulting in each case.

In each configuration, issues related to item sequence are also discussed. As indicated in Chapter III, five item sequences were randomly selected from the 48 unique alternating sequences presented in Table 6. These five sequences were used in the evaluation of each case, and the sequence resulting in the lowest total inventory holding cost was considered the near-optimum (in some cases, clearly optimal) sequence.

The five item sequences and their respective alphabetical codes are shown below:

Item Sequence	Code
1-2-3-4-5-6-7-8	A
1-2-8-3-5-6-4-7	B
1-4-6-3-5-8-2-7	C
1-6-4-7-5-2-8-3	D
1-4-2-3-5-8-6-7	E

Configuration I

System Overview

Figure 11 shows the structure of system configuration I and the paths followed by all eight items through the system. The bottleneck work center (B) is at the front-end of the system, feeding work centers C1, C2, C3, and C4. The base case parameters of the system under this configuration are shown in Table 13. As explained in Chapter III, these parameters have been selected such that the bottleneck work center is the determinant of system capacity in the totally conventional system.

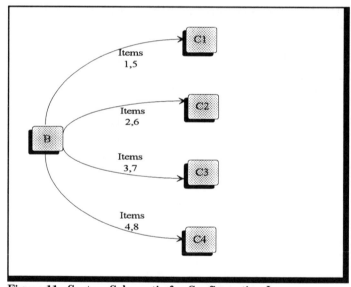

Figure 11. System Schematic for Configuration I

Chapter 4: Results 63

Table 13
System Parameters for Configuration I

Paramete	1	2	3	4	5	6	7	8
System								
Demand	700	350	900	970	30	600	300	690
Work Center B								
S	4.50	3.50	2.750	2.250	3.0	5.00	2.250	0.500
p	0.03	0.02	0.040	0.040	0.0	0.03	0.050	0.050
h	0.02	0.03	0.040	0.060	0.0	0.08	0.090	0.170
Work Center C1								
S	9.11	...	...	...	9.2	...	...	...
p	0.05	...	...	...	0.2	...	...	...
h	0.03	...	...	...	0.1	...	...	...
Work Center C2								
S	...	7.02	...	...	...	5.83	...	...
p	...	0.17	...	...	...	0.06	...	...
h	...	0.08	...	...	...	0.11	...	...
Work Center C3								
S	...	...	1.00	...	...	...	1.58	...
p	...	...	0.10	...	...	...	0.35	...
h	...	...	0.08	...	...	...	0.17	...
Work Center C4								
S	...	...	...	9.26	...	...	...	9.62
p	...	...	...	0.11	...	...	...	0.11
h	...	...	...	0.08	...	...	...	0.23

Legends:
  S = Setup times (days)
  p = Production times (per unit)
  h = Holding cost (dollar per unit per year)

Flexible Manufacturing Islands 64

Case 1: Conventional System

The conventional system involves no FMI and uses the system parameters presented in Table 13. The procedure for projecting the inventory behavior in each configuration or case, is the same as the procedure discussed in Chapter III for the simplified example. However, the procedure is briefly reviewed here. First, the minimum feasible cycle durations (C_{min}) were calculated for each work center using Equation 1. The C_{min} values were checked to ensure that the bottleneck work center produced the largest value. In this case, the C_{min} for work center B is 48.11 days, which is the minimum feasible (and therefore optimum) cycle duration for the system. Then, lots sizes for all items were calculated as the demand rate multiplied by the system cycle duration. The production cycle for work center B and the inventory following the work center were then projected as illustrated in Table 14 for item sequence A. The production schedule and inventory behavior following work center C1 for items #1 and #5 in the same sequence are shown in Table 15.

Table 14
Production Schedule and Inventory Behavior Following Work Center B
Configuration I, Conventional System, Base Case, Sequence A

Item	Lot Size (Units)	Start Setup (Days)	Start Prod. (Days)	End Prod. (Days)	Peak Amount (Units)	Start Cons.* (Days)	End Cons.* (Days)
1	93.56	0.00	4.50	7.31	93.56	7.31	11.98
2	46.78	7.31	10.81	11.74	46.78	46.78	19.69
3	120.29	11.74	14.49	19.30	120.29	19.30	31.33
4	129.64	19.30	21.55	26.74	129.64	26.74	41.00
5	40.10	26.74	29.74	31.34	40.10	31.34	41.37
6	80.19	31.34	36.34	38.75	80.19	38.75	43.56
7	40.10	38.75	41.00	43.00	40.10	43.00	57.04
8	92.22	43.00	43.50	48.11	92.22	48.11	58.26

* Start and end of consumption.

Chapter 4: Results 65

Table 15
Inventory Behavior Following Work Center C1:
Configuration I, Conventional System, Base Case, Sequence A

Item	Lot Size (Units)	Start Setup (Days)	Start Prod. (Days)	End Prod. (Days)	Peak Amount (Units)	Start Cons.* (Days)
1	-1.80	7.31	11.98	84.46	11.98	55.42
5	22.12	31.34	41.37	31.74	41.37	79.46

* Start and end of consumption.

Setup for item #1 begins at the beginning of the cycle (day 0.00) and continues for 4.5 days, at which time production begins. It takes 2.81 days to produce the lot size of 93.56 units (at 0.03 days per unit) of item #1 at work center B, and production ends at day 7.31. The inventory builds up at the rate of production for each item at work center B until it reaches the peak amount, which is equal to the lot size. Thus, the inventory of item #1 builds up at the rate of 33.33 units per day (i.e., 1/.03) to a peak of 93.56 units. At this point, consumption begins and the inventory is depleted at a rate equal to the production rate for item #1 at work center C1, which is 20 units per day. Therefore, the inventory level for item #1 following work center B reaches zero on day 11.98. The setup for item #2 begins after production of item #1 has ended (day 7.31). The calculations for all other items are similar to item #1. Note that the end of production for item #8 occurs at day 48.11, which is the end of the cycle.

The scheduling of items processed in work center C1 (see Table 15) is tied to the schedule in work center B. Thus, the end of production for items #1 and #5 at work center B determines the start of production for the same items at work center C1. The start of setup for each item is calculated by subtracting the setup time from the start of production. The end of production is based on the start of production plus the time required to produce the lot. Inventory following work center C1 accumulates at a rate equal to the production rate minus the demand rate for the item. Thus, the peak inventory levels are less than the lot sizes. The start of consumption in this case actually refers to the time that the inventory begins to decline, which is at the end of production. The end of consumption refers to the time that the inventory reaches zero, which is calculated as the start of consumption plus the time required to consume the peak inventory amount at the demand rate for the item. Note that for both items, the length of time between the start of production and the end of consumption is equal to the cycle duration (48.11 days). In other words, the inventory following work center C1 reaches zero just when it is time to start producing the item again in the next cycle. Also, note that the length of time between the start of setup for item #1 (day -1.80) and the end of production for item #5 (day 41.37) is 43.17 days, which is less than the system cycle duration.

Finally, it should be noted that no schedule conflict occurs in this case in work center C1. This observation is based on the fact that the start of setup for item #5 is later than the end of production for item #1. If this had not been true, the start of setup for item #5 would have been delayed until production of item #1 was finished.

The analysis for work center C2, C3, and C4 followed the same logic as explained above for work center C1. The calculation of inventory holding cost following each work center, simply involves the calculation of the areas associated with the inventory behavior for each item. These areas are then multiplied by the holding cost parameter for the items to determine the inventory holding cost for one cycle. Finally, the inventory holding cost for the cycle duration is annualized based on 360 days per year.

Base Case

The five item sequences were evaluated to determine which sequence resulted in the lowest inventory holding cost. Three of the item sequences led to schedule conflicts in work center C4. As shown in Chapter III (see Figure 5), when schedule

Chapter 4: Results 67

conflict occurs, it requires inventory for items involved in the conflict to be held for a longer period of time than would be necessary if the conflict had not occurred. This increases the inventory holding cost for the system. Consequently, an item sequence which does not result in schedule conflict in any work center is clearly an optimal item sequence, since it minimizes the inventory holding cost. In the base case for configuration I, item sequences C and E caused no schedule conflict in any work center and, therefore, are clearly optimum item sequences for that case. Both sequences resulted in the inventory holding costs shown in Table 16.

Sensitivity Analysis

As indicated in Chapter III, the sensitivity analysis was conducted at two levels, high and low (taking the base case as the medium level), for the following three sets of system parameters:

1. Capacity utilization rate
2. Setup time to run time ratio
3. Rate of increase in inventory holding cost rates

The methods for manipulating the system parameters were explained in Chapter III, and the resulting parameters are presented in Appendix A. As in the base case, item sequences C and E caused no schedule conflict in any of the sensitivity analyses and were, therefore, clearly optimum. The annual inventory holding costs for each work center and for the total system are presented in Table 17 for each of the sensitivity analyses. Table 18 compares the annual system holding costs for each sensitivity analysis to the base case. In addition to presenting the costs as dollar amounts, Table 18 expresses each cost as a multiple of the base case cost.

Flexible Manufacturing Islands 68

Table 16
Annual Inventory Holding Cost:
Configuration I, Conventional System
Base Case, Sequence C or E

Work Center	Inventory Cost
B	$ 2,336
C1	1,085
C2	1,991
C3	2,959
C4	5,685
Total	$14,056

Table 17
Total Inventory Holding Cost Under Sensitivity Analyses
Configuration I, Conventional System, Sequence C or E

Work Center	Capacity Utilization		Setup/Run Time		Rate of Holding Cost	
	Low	High	Low	High	Low	High
B	$875	$5,702	$1,635	$3,036	$ 2,686	$1,985
C1	611	1,886	751	1,394	980	1,198
C2	1,109	3,587	1,394	2,589	1,807	2,182
C3	1,584	5,557	2,071	3,847	2,273	3,186
C4	3,403	10,676	3,979	7,390	5,051	6,318
Total	$7,222	$27,408	$9,830	$18,256	$13,256	$14,860

Chapter 4: Results 69

Table 18
Comparison of Inventory Holding Costs Under the Base Case and the Sensitivity Analyses
Configuration I, Conventional System, Sequence C or E

Level	Capacity Utilization	Setup/Run Time Ratio	Rate of Increase in Inventory Cost
High	$27,408 (1.95)[a]	$18,256 (1.30)	$14,860 (1.06)
Base	$14,056 (1.00)		
Low	$7,222 (0.51)	$9,830 (0.70)	$13,256 (0.94)

[a] Figures in parentheses express the cost as a multiple of the base case cost.

Case 2: Work Center B is FMI

Base Case

In this case, work center B, the bottleneck work center, was considered to be the FMI. As indicated in Chapter III, the primary distinguishing characteristic of an FMI work center is that the setup times for all items processed in the work center are zero. Therefore, in case 2, setup times for all items at work center B are zero. Other parameters of the system remain the same as in the conventional system case (see Table 14).

Although when considered an FMI, work center B could operate in unit flow mode, this work center in configuration I still operates in batch mode, because the following work centers (i.e. C1, C2, C3, and C4) are all still operating in batch mode.

The cycle duration for the system is no longer determined by work center B, since setup times at that work center are zero. Instead, the system cycle duration is the largest of the C_{min} values for the non-bottleneck work centers. In this case, the C_{min} for work center C4 is the largest (38.31 days). As a result of the shorter cycle duration, the lot sizes for all items decrease as compared to the conventional system.

Since work center C4 now determines the cycle duration for the system, the process of projecting the production cycle started at work center C4. The production start times for items #4 and #8 at work center C4 established the production ending times for these items at work center B from which the production start times for these items at work center B were derived. The remaining items were scheduled around items #4 and #8 in work center B according to the item sequence being used. The production ending times for these items at work center B in turn determined their production starting times at work centers C1, C2, and C3.

The five item sequences were evaluated for this case. Again, item sequences C and E caused no conflict and, therefore, were the optimal sequences. Both sequences resulted in the inventory holding costs shown in Table 19. As compared to the base case for the conventional system, the holding cost in this case are lower for all work centers primarily because of the smaller lot sizes, which are made possible by shorter cycle duration.

Sensitivity Analysis

The parameters used in the sensitivity analysis are the same as for the conventional system (see Appendix A), except that the setup times for all items at work center B are zero. Again, item sequences C and E produced no schedule conflict in any of the sensitivity analyses and were, therefore, clearly optimal.

The annualized inventory holding costs for the various sensitivity analyses are shown in Table 20. Table 21 compares the annual inventory holding costs for each sensitivity analysis to the base case.

Chapter 4: Results

Table 19
System Inventory Holding Cost:
Configuration I, B is FMI, Base Case

Work Center	Cost
B	$996
C1	864
C2	1,586
C3	1,726
C4	3,445
Total	$8,617

Table 20
Inventory Holding Cost Under Sensitivity Analyses: Configuration I, B is FMI, Sequence C or E

Work Center	Capacity Utilization		Setup/Run Time		Rate of Holding Cost	
	Low	High	Low	High	Low	High
B	$ 417	$2,644	$488	$1,684	$1,146	$ 847
C1	1489	1,528	605	1,123	780	974
C2	2883	2,861	1,110	2,061	1,335	1,562
C3	1,024	2,892	1,209	2,244	1,594	1,859
C4	2,01	6 5,86	92,412	4,479	3,061	3,829
Total	$4,829	$15,794	$5,8241	$1,591	$7,916	$9,071

Table 21
Comparison of Inventory Holding Costs Under the Base Case and the
Sensitivity Analyses: Configuration I, B is FMI, Sequence C or E

Level	Capacity Utilization	Setup/Run Time	Rate of Increase in Inventory Holding Cost
High	$15,794 (1.83)[a]	$11,591 (1.35)	$9,071 (1.05)
Base		$8,617 (1.00)	
Low	$4,829 (0.56)	$5,824 (0.68)	$7,916 (0.92)

[a] Figures in parentheses express the cost as a multiple of the base case cost.

Case 3: Work Center C4 is FMI

Base Case

In this phase of the study, a non-bottleneck work center was converted into an FMI. In determining which non-bottleneck work center to convert, total inventory holding costs for all such work centers in the base case for the conventional system were compared (see Table 16). Since work center C4 had the highest cost, it was selected for conversion. The system parameters are the same as for the conventional system, except that the setup times for items #4 and #8 at work center C4 are zero.

Since work center B is again a conventional work center, it determines the system cycle duration, which is the same as it was in the base case for the conventional system (48.11 days). Also, the analysis for this case began with the projection of the production cycle for work center B, as it did for the conventional system.

Unit flow is now possible from work center B to C4. However, because work center B operates in batch mode, C4 must operate in batch mode. Therefore, inventory builds up for

Chapter 4: Results

items #8 and #4 between work center B and C4. The rate of inventory build-up, however, is not equal to the rate of production for these items at work center B (as it is for items 1, 2, 3, 5, and 7). The inventory following work center B for items #4 and #8 builds up at a rate equal to the production rate at work center B minus the demand rate for the item. In other words, work center C4 processes the items unit-by-unit at rates just adequate to meet demand. The inventory peaks for items #4 and #8 (115.67 and 83.38 units respectively) following work center B are smaller than the lot sizes for these items. These peaks are equal to the build-up rate as explained above multiplied by the time required to produce the entire lot at work center B. Since the inventory following work center B for items #4 and #8 is depleted at the demand rates (rather than the production rates at work center C4) for those items, the inventory is not fully depleted until it is time to start producing the items again in the next cycle. This significantly increases the holding cost for these items at work center B. Of course, there is no inventory build-up at all following work center C4 in this case.

Of the five items sequences, only sequence C produced no schedule conflict in this case. The results of the analysis based on sequence C are shown in Table 22. Note that the holding costs for work centers C1, C2, C3, and C3 are the same as in the case for the conventional system (see Table 16). The holding cost at work center B, however, has increased and the cost at work center C4 has become zero for the reasons explained above.

Sensitivity Analysis

The parameters used in the sensitivity analysis are the same as for the conventional system (see Appendix A), except that the setup times for items #4 and #8 at work center C4 are zero.

Again, item sequence C caused no schedule conflict in any of the sensitivity analyses. The results of the sensitivity analysis are presented in Table 23. Table 24 compares the

Flexible Manufacturing Islands 74

annual inventory holding costs for each sensitivity analysis to the base case.

Table 22
Annual Inventory Holding Cost:
Configuration I, Base Case, C4 is FMI Sequence C

Work Center	Cost
B	$4,705
C1	1,085
C2	1,991
C3	2,959
C4	0
Total	$10,740

Table 23
Inventory Holding Cost Under Sensitivity Analyses: Configuration I, C4 is FMI, Sequence C

Work Center	Capacity Utilization		Setup/Run Time		Rate of Holding Cost	
	Low	High	Low	High	Low	High
B	$2,441	$9,108	$3,294	$6,117	$5,411	$3,999
C1	611	1,886	751	1,394	980	1,189
C2	1,109	3,587	1,394	2,589	1,807	2,182
C3	1,584	5,557	2,071	3,847	2,732	3,186
C4	0	0	0	0	0	0
Total	$5,745	$20,138	$7,510	$13,947	$10,930	$10,556

Table 24
Comparison of Inventory Holding Costs Under the Base Case
and the Sensitivity Analyses:

Level	Capacity Utilization	Setup/Run Time Ratio	Rate of Increase in Inventory Cost
High	$20,138 (1.88) [a]	$13,947 (1.30)	$10,556 (0.98)
Base		$10,740 (1.00)	
Low	$5,745 (0.53)	$7,510 (0.70)	$10,930 (1.02)

[a] Figures in parentheses express the cost as a multiple of the base case cost.

In Brief

The results of the analysis for configuration I are summarized in Table 25. The annual inventory holding cost is shown for the conventional system and for both positions of the FMI under the base case and under the various sensitivity analyses. The figures in parentheses express the cost for that case as a multiple (or percentage) of the cost for the conventional ("No FMI") case in that same column. Figures 12, 13, and 14 graphically show the comparison of the holding costs presented in Table 25.

Table 25
Inventory Holding Cost Under Various Cases and FMI Positions:
Configuration I

FMI Position	Base Case	Capacity Utilization		Setup/Run Time Ratio		Rate of Inventory Holding Cost Increase	
		Low	High	Low	High	Low	High
No FMI	$14,056 (1.00) *	$7,222 (1.00)	$27,408 (1.00)	$9,830 (1.00)	$18,256 (1.00)	$13,256 (1.00)	$14,860 (1.00)
B FMI	8,617 (0.61)	4,829 (0.58)	15,794 (0.67)	5,824 (0.63)	11,591 (0.59)	7,916 (0.61)	9,071 (0.60)
C4 FMI	10,740 (0.76)	5,745 (0.73)	20,138 (0.80)	7,510 (0.76)	13,947 (0.76)	10,930 0.71)	10,556 (0.82)

* Figures in parentheses express the cost as a multiple of the base case cost.

Chapter 4: Results

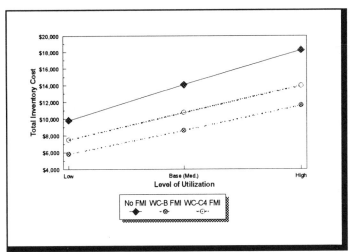

Figure 12. Comparison of Inventory Holding Cost: Setup/Run Time Rations vs. FMI Position in Configuration I

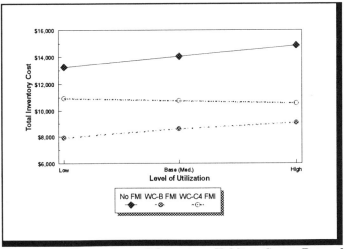

Figure 13. Comparison of Inventory Holding Cost: Rate of Holding Cost Increase vs. FMI Position in Configuration I

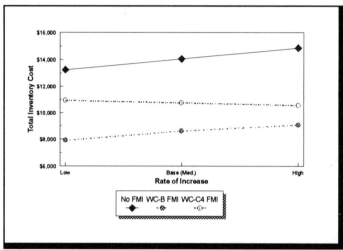

Figure 14. Comparison of Inventory Holding Cost: Rate of Inventory Holding Cost Increase vs. FMI Position in Configuration I

Chapter 4: Results 79

Configuration III

System Overview

Configuration III is very similar in structure to the simplified example discussed in Chapter III. The major distinctions are that in configuration III there are eight items and four non-bottleneck work centers, while in the simplified example there were only four items and two non-bottleneck work centers. Figure 15 shows the system schematic for configuration III

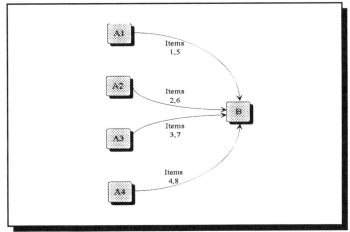

Figure 15. System Schematic for Configuration III

The base case parameters of the system in this configuration are given in Table 26. The annual demand values for the items are the same as in configuration I, as are the setup time and run time parameters for the bottleneck work center. The setup and run time parameters for the non-bottleneck work centers were set such that the minimum cycle duration (C_{min}) values for these work centers are in the same range as they were for the non-bottleneck work centers in configuration I.

Chapter 4: Results

Table 26
System Parameters for Configuration III

Parameters	Items							
	1	2	3	4	5	6	7	8
System								
Demand	700	350	900	970	300	600	300	690
Work Center A1								
S	9.11	.....	.....	.....	9.22	.....	.....	.....
p	.05	.....	.....	.....	.25	.....	.....	.....
h	.02	.....	.....	.....	.06	.....	.....	.....
Work Center A2								
S	.....	7.02	.....	.....	.....	5.83	.....	.....
p	.....	.17	.....	.....	.....	.06	.....	.....
h	.....	.03	.....	.....	.....	.08	.....	.....
Work Center A3								
S	.....	.....	1.00	.....	.....	.....	1.58	.....
p	.....	.....	.10	.....	.....	.....	.35	.....
h	.....	.....	.04	.....	.....	.....	.09	.....
Work Center A4								
S	.....	.....	.....	9.26	.....	.....	.....	9.26
p	.....	.....	.....	.11	.....	.....	.....	.11
h	.....	.....	.....	.06	.....	.....	.....	.17
Work Center B								
S	4.50	3.5	2.75	2.25	3.00	5.00	2.25	.50
p	.03	.02	.04	.04	.04	.03	.05	.05
h	.03	.04	.08	.08	.11	.17	.17	.23

Legends:
    S =   Setup times (days)
    p =   Production times (per unit)
    h =   Holding cost (dollar per unit per year)

Flexible Manufacturing Islands 82

Case 1: Conventional System

The conventional system involves no FMI and uses the system parameters presented in Table 26. The procedure for projecting the inventory behavior in this configuration is exactly the same as the one described in Chapter III for the simplified example. The procedure is also very similar to configuration I, except that in configuration III, the non-bottleneck work center is at the back end of the system. Table 27 shows the production cycle for work center B and the inventory following the work center for item sequence A.

Table 27
Production Schedule and Inventory Behavior Following Work Center B
Configuration III, Conventional System, Base Case, Sequence A

Item	Lot Size	Start Setup	Start Prod.	End Prod.	Peak Amount	Start Cons.*	End Cons.*
1	93.56	0.00	4.50	7.31	88.10	7.31	52.61
2	46.78	7.31	10.81	11.74	45.87	46.78	58.92
3	120.29	11.74	14.49	19.30	108.26	19.30	62.61
4	129.64	19.30	21.55	26.74	115.67	26.74	69.67
5	40.10	26.74	29.74	31.34	38.76	31.34	77.85
6	80.19	31.34	36.34	38.75	76.18	38.75	84.46
7	40.10	38.75	41.00	43.00	38.43	43.00	89.11
8	92.22	43.00	43.50	48.11	83.38	48.11	91.62

*Start and end of consumption.

Note that the "Start Setup", "Start Production", "End Production", and "Start Consumption" values are all the same as in configuration I. However, due to the fact that work center B is now at the back end of the system and the consumption rate for all items at this work center is equal to their demand rate, the inventory peak amounts are not equal to lot sizes. Instead, they are equal to the difference between the production rate and the demand rate for each item multiplied by the time required to produce the lot. Also, since

Chapter 4: Results

the consumption rate is equal to the demand rate for each item, the "End of Consumption" is different from the same figures in configuration I.

The scheduling of items processed at work centers A1, A2, A3, and A4 is tied to the schedule in work center B. Thus, the start of production at work center B for item #1 and #5, as an example, determines the end of production for these items at work center A1. In turn, the end of production for these items at work center A1 determines their start of setup times at this work center. Table 28 shows the production schedule and inventory behavior following work center A1 for items #1 and #5 for item sequence A. Note that the peak amounts for item #1 and #5 at work center A1 are equal to their respective lot sizes. Note also that no schedule conflict occurs in this case between the end of production for item #1 and the start of setup for item #5. Finally, the length of time between the start of setup for item #1 and the end of production for item #5 is less than the system cycle duration. The analysis for work centers A2, A3, and A4 followed the same logic as explained above for work center A1. The calculation of inventory holding cost following each work center simply involves the calculation of the areas associated with the inventory behavior for each item. These areas are then multiplied by the holding cost parameter for the items to determine the inventory holding cost for one cycle. Finally, the inventory holding cost for the cycle duration is annualized based on 360 days per year.

Table 28
Production Schedule and Inventory Behavior Following Work Center A1
Configuration III, Conventional System, Base Case, Sequence A

Item	Start Setup (Days)	Start Prod. (Days)	End Prod. (Days)	Peak Amount (Units)	Start Cons.* (Days)	End Cons.* (Days)
1	-9.29	-0.18	4.50	93.56	4.50	7.31
5	10.50	19.72	29.74	40.10	29.74	31.34

* Start and end of consumption.

Base Case

The five item sequences were evaluated to determine which sequence resulted in the lowest inventory holding cost. None of the item sequences resulted in schedule conflict in this case. Thus, all sequences were optimal and all sequences resulted in inventory holding costs shown in Table 29.

Table 29
Annual Inventory Holding Cost
Configuration III, Conventional System
Base Case, All Five Sequences

Work Center	Inventory Cost
A1	$165
A2	220
A3	519
A4	1,431
B	11,757
Total	$14,092

Sensitivity Analysis

Sensitivity analyses were carried out in the same manner as in configuration I. Similarly, the parameters of the system were manipulated as explained in Chapter III and the resulting parameters are presented in Appendix B.

The five item sequences were again evaluated to determine which resulted in the lowest inventory holding cost for the system. Of the five item sequences, sequences A, B, and D resulted in schedule conflict under certain sensitivity analyses at work center A4, while sequences C and E did not produce conflict in any of the sensitivity analyses in any work center. Therefore, item sequences C and E were clearly the optimal sequences. The annual inventory holding costs for each work

Chapter 4: Results 85

center and for the total system are presented in Table 30 for each of the sensitivity analyses. Table 31 compares the annual system holding costs for each sensitivity analysis to the base case. In addition to presenting the costs as dollar amounts, the table expresses each cost as a multiple of the base case cost.

Table 30
Total Inventory Holding Cost Under Sensitivity Analyses
Configuration III, Conventional System, Sequence C or E

Work Center	Capacity Utilization		Setup/Run Time		Rate of Holding Cost	
	Low	High	Low	High	Low	High
B	$6,450	$21,529	$8,230	$15,284	$10,657	$12,857
A1	62	403	115	214	190	140
A2	82	537	154	286	253	187
A3	268	1,744	500	929	823	607
A4	539	3,494	1,002	1861	1646	1,217
Total	$7,398	$27,707	$10,001	$18,574	$13,569	$15,008

Table 31
Comparison of Inventory Holding Costs Under the Base Case and the Sensitivity Analyses
Configuration III, Conventional System, Sequence C or E

Level	Capacity Utilization	Setup/Run Time Ratio	Rate of Increase in Inventory Cost
High	$2,7707 (1.97)	$18,574 (1.32)	$15,008 (1.07)
Base		14,092 (1.00)	
Low	$7,398 (0.52)	$10,001 (0.71)	$1,3569 (0.96)

[a] Figures in parentheses express the cost as a multiple of the base case cost.

Flexible Manufacturing Islands 86

Base Case
Case 2: Work Center B is FMI

The primary distinguishing characteristic of the system under this case, as compared to the conventional system, is that the setup times for all items in work center B are zero. All other parameters of the system remain the same as for the conventional system.

Once work center B is converted into an FMI, it ceases to be the bottleneck work center. The effect of this change is that the non-bottleneck work centers A1, A2, A3, and A4 become independent with respect to cycle duration. In other words, when work center B is converted into an FMI there is no single cycle duration for the entire system. Instead, each non-bottleneck work center operates on its own minimum feasible cycle duration. These cycle durations are 26.40 days, 17.49 days, 5.63 days, and 38.31 days for work centers A1, A2, A3 and A4 respectively. Note that the longest C_{min} is shorter in duration than the system cycle duration (48.11 days) when the system was totally conventional.

Since there is no single cycle duration for the system, and all non-bottleneck work centers are independent of the bottleneck work center, clearly there can be no schedule conflict in any work center. Therefore, the issue of item sequence becomes irrelevant. Under these circumstances, the setup for the first item in each non-bottleneck work center starts at day zero. After setting up for the first item, production of the item starts. After the production of the first item, setup for the second item begins an so on. To illustrate, the production schedule and inventory behavior for work center A1 is shown in Table 32.

Note that the peak amounts for item #1 and #5 are smaller than their respective lot sizes. These amounts are equal to the difference between the production rate for the item at work center A1 and its demand rate multiplied by the time required to produce the lot. The peak inventory amounts are depleted

Chapter 4: Results 87

at the demand rates of the items, since work center processes the items unit-by-unit at rates just adequate to meet demand. Table 33 shows the system inventory holding cost for configuration III, under the base case, when work center B is an FMI.

Table 32
Production Schedule and Inventory Behavior Following Work Center A1
Configuration III, B is FMI, Base Case, All Sequences

Item	Lot Size (Units)	Start Setup (Days)	Start Prod. (Days)	End Prod. (Days)	Peak Amount (Units)	Start Cons.* (Days)	End Cons.* (Days)
1	93.56	0.00	4.50	7.31	88.10	7.31	52.61
2	46.78	7.31	10.81	11.74	45.87	46.78	58.92
3	120.29	11.74	14.49	19.30	108.26	19.30	62.61
4	129.64	19.30	21.55	26.74	115.67	26.74	69.67
5	40.10	26.74	29.74	31.34	38.76	31.34	77.85
6	80.19	31.34	36.34	38.75	76.18	38.75	84.46
7	40.10	38.75	41.00	43.00	38.43	43.00	89.11
8	92.22	43.00	43.50	48.11	83.38	48.11	91.62

* Start and end of consumption.

Sensitivity Analysis

The parameters used in the sensitivity analysis are the same as for the conventional system (see Appendix B), except that the setup times for all items at work center B are zero. The annualized inventory holding costs for the various sensitivity analyses are shown in Table 34. Table 35 compares the annual inventory holding costs for each sensitivity analysis to the base case.

Flexible Manufacturing Islands 88

Sensitivity Analysis

The parameters used in the sensitivity analysis are the same as for the conventional system (see Appendix B), except that the setup times for all items at work center B are zero. The annualized inventory holding costs for the various sensitivity analyses are shown in Table 34. Table 35 compares the annual inventory holding costs for each sensitivity analysis to the base case.

Table 33
Annual Inventory Holding Cost
Configuration III, B is FMI
Base Case, All Five Sequences

Work Center	Inventory Cost
A1	$380
A2	454
A3	177
A4	2,558
B	0
Total	$3,569

Table 34
Total Inventory Holding Cost Under Sensitivity Analyses
Configuration III, Conventional System, Sequence C or E

Work Center	Capacity Utilization		Setup/Run Time		Rate of Holding Cost	
	Low	High	Low	High	Low	High
A1	$248	$537	$266	$494	$437	$323
A2	$298	$638	$318	$591	$601	$444
A3	$92	$357	$124	$231	$204	$151
A4	$1,497	$4,357	$1,790	$3,325	$2,941	$2,174
B	$0	$0	$0	$0	$0	$0
Total	$2,135	$5,889	$2,498	$4,641	$4,183	$3,092

Table 35
Comparison of Inventory Holding Costs Under the Base Case and the
Sensitivity Analyses
Configuration III, B is FMI, All Sequences

Level	Capacity Utilization	Setup/Run Time Ratio	Rate of Increase in Inventory Cost
High	$5,889 (1.65)	$4,641 (1.30)	$3,092 (0.87)
Base		3569 (1.00)	
Low	$2,135 (0.60)	$2,498 (0.70)	$4,183 (1.17)

[a] Figures in parentheses express the cost as a multiple of the base case cost.

Chapter 4: Results 91

Case 3: Work Center A4 is FMI

Base Case

In this phase of the analysis, work center A4 was converted into an FMI. The decision to convert this work center was based on the fact that the inventory holding cost for this work center (see Table 29) was the highest of all non-bottleneck work centers in the base case for the conventional system. The setup times for items #4 and #8 are zero. All other parameters are the same as for the conventional system.

Since, work center B is conventional in this case, its C_{min} (48.11 days) establishes the cycle duration for the system. The scheduling of the non-bottleneck work centers is tied to the schedule in work center B. In this case, the start of production for a given item at work center B determines the end of production for the same item at the non-bottleneck work centers which processes that item.

As an FMI, work center A4 operates in a unit flow mode. However, the production rates for items #4 and #8 in work center B are faster than the production rates for these items at work center A4. Therefore, it is necessary to build up some inventory of items #4 and #8 between work centers A4 and B before work center B starts production of these items. Without this initial inventory build-up following work center A4, work center B would be starved. The amount of inventory build-up following work center A1 for items #4 and #8 was determined using Equation 5 (in Chapter III) to be 82.50 units for item #4 and 50.30 units for item #8. These amounts were accumulated at the rate of 9.09 units per day, which is the production rate for both items at work center A4. The peak inventory amounts for items #4 and #8 following work center A4 are depleted at net rates of 15.91 and 19.91 units per day respectively. The depletion of the inventory for each item begins when work center B begins to process the item, and these rates of consumption are equal to the difference between the processing rates for these items at work centers B and A4. Table 36

Flexible Manufacturing Islands 92

shows the production schedule and inventory behavior for work center A4 after it has been converted into an FMI.

Table 36
Production Schedule and Inventory Behavior Following Work Center A4 Configuration III, A4 is FMI, Base Case, Sequence A

Item	Start Setup (Days)	Start Prod. (Days)	End Prod. (Days)	Peak Amount (Units)	Start Cons.[a] (Days)	End Cons.[a] (Days)
4	N/A	12.48	21.55	82.50	21.55	26.74
8	N/A	41.10	43.50	50.30	43.50	48.11

[a] Start and end of consumption.

The production schedule and inventory behavior for work center B and the non-bottleneck work centers A1, A2, and A3 were similar to the case when the system was totally conventional.

As mentioned in case 1 (the conventional system), schedule conflict occurred at work center A4 for item sequences A, B, and D in some of the sensitivity analyses. Once this work center was converted into an FMI, none of the five item sequences caused any conflict in any work center under any case. Therefore, all item sequences were optimal. Table 37 shows the inventory holding cost for in configuration III when work center A4 is converted into an FMI. Notice that the inventory holding costs for work centers A1, A2, A3, and B are the same as in the conventional case, since converting work center A4 into an FMI does not affect their inventory behavior.

Chapter 4: Results 93

Table 37
Annual Inventory Holding Cost
Configuration III, Conventional System
Base Case, A4 is FMI, All Sequences

Work Center	Inventory Cost
A1	$165
A2	220
A3	519
A4	589
B	11,757
Total	$13,250

Sensitivity Analysis

The parameters used in the sensitivity analysis are the same as for the conventional system (see Appendix B), except that the setup times for items #4 and #8 at work center A4 are zero. The annualized inventory holding costs for the various sensitivity analyses are shown in Table 38. Table 39 compares the annual inventory holding costs for each sensitivity analysis to the base case.

Table 38
Total Inventory Holding Cost Under Sensitivity Analyses
Configuration III, A4 is FMI, All Sequences

Work Center	Capacity Utilization		Setup/Run Time		Rate of Holding Cost	
	Low	High	Low	High	Low	High
A1	$62	$403	$115	$214	$190	$140
A2	82	537	154	286	253	187
A3	268	1,744	500	929	823	607
A4	221	1,437	412	765	678	501
B	6,450	21,529	8,230	15,284	10,657	12,857
Total	$7,083	$25,650	$9,411	$17,478	$12,601	$14,292

Table 39
Comparison of Inventory Holding Costs Under the Base Case and the
Sensitivity Analyses: Configuration III, A4 is FMI, All Sequences

Level	Capacity Utilization	Setup/Run Time Ratio	Rate of Increase in Inventory Cost
High	$25,650 (1.94)	$17,478 (1.32)	$14,292 (1.08)
Base	$13,250 (1.00)		
Low	$7,083 (0.53)	$9,411 (0.71)	$12,601 (0.95)

a Figures in parentheses express the cost as a multiple of the base case cost.

Chapter 4: Results

In Brief

The results of the analysis for configuration III are summarized in Table 40. The annual inventory holding cost is shown for the conventional system and for both positions of the FMI under the base case and under various sensitivity analyses. The figures in parentheses express the cost for that case as a multiple (or percentage) of the cost for the conventional ("No FMI") case in that same column. Figures 16, 17, and 18 graphically show the comparison of the holding costs presented in Table 40.

Table 40
Total Inventory Holding Cost Under Various Cases and FMI Positions: Configuration III

FMI Position	Base Case	Capacity Utilization		Setup/Run Time		Rate of Holding Cost	
		Low	High	Low	High	Low	High
No FMI	$14,092 (1.00)	$7,398 (1.00)	$27,707 (1.00)	$10,001 (1.00)	$18,316 (1.00)	$13,569 (1.00)	$15,008 (1.00)
B FMI	3,569 (0.25)	2,135 (0.29)	5,889 (0.21)	2,498 (0.25)	4,641 (0.25)	4,183 (0.31)	3,092 (0.23)
A4 FMI	13,250 (0.25)	7,083 (0.96)	25,650 (0.94)	9,411 (0.94)	17,478 (0.95)	12,601 (0.93)	14,292 (1.05)

[a] Figures in parentheses express the cost as a multiple of the "No FMI " cost.

The following conclusions about configuration III can be drawn from the results of configuration III presented here and from Table 40 and Figures 17, 18, and 19:

1. Converting a work center into an FMI reduces inventory holding cost of the system, regardless of the position of the work center in the system.
2. Reduction in system's inventory holding cost is greatest when the work center converted into an FMI is the bottleneck work center (B).

Flexible Manufacturing Islands 96

3. In this configuration, converting the bottleneck work center (B) into an FMI resulted in a relatively larger reduction in inventory cost than in configuration I. The primary reason for this was that inventory of items following work center B in this configuration is consumed over a longer period of time than in Configuration I (because the rates of consumption are slower, equals demand, in this configuration). Therefore, when work center B is converted into an FMI, it eliminates the inventory following work center B and significantly reduces inventory holding cost of the system. Furthermore, work center B (being at the back end of the system) has higher holding cost rates in this configuration that in Configuration I.
4. Converting a non-bottleneck work center (A4) into an FMI in this configuration had significantly smaller impact on reducing inventory cost than in configuration I. The reason was, again, the fact that in Configuration III this work center was located at the front-end of the system, with relatively low per unit inventory holding cost. Furthermore, the FMI must still operate in batch mode and inventory must be built up between work centers A4 and B to prevent starvation of work center B.
5. Inventory holding cost seems to be highly sensitive to the changes in capacity utilization rate, in the same manner as in configuration I and similarly, the relationship between inventory cost and capacity utilization is not linear.
6. Inventory holding cost seems to be linearly related to setup time to run time ratio for all positions of FMI within the range of sensitivity analysis conducted. This is clearly notable from Tables 31, 35, and 39. These tables show that 30 percent increase (decrease) in the setup time to run time ratio resulted in 30 percent increase (decrease) in inventory holding cost for the system.
7. Inventory holding cost seems to be relatively insensitive to the rate of increase in holding cost.

Chapter 4: Results 97

8. The relationship between inventory holding cost and rate of increase in holing cost is approximately linear. Furthermore, this relationship is positive in all cases except when bottleneck work center (B) is converted into an FMI. When the bottleneck work center is converted into an FMI, the relationship becomes negative. The explanation for this negative relationship is as follows. When work center B is converted into an FMI, the inventory for items going through this work center will be zero and when the rate of increase in holding cost is manipulated upward (in "low rate of increase in holding cost" case), the inventory holding cost for the non-bottleneck work centers increase while there is no change in inventory holding cost for work center B (remains zero). This results in higher total inventory holding cost under "low rate of increase" in inventory holding cost. On the other hand, in the "high rate of increase" in holding cost, the total inventory holding cost for the system decreases because the holding cost per units for all items are decreased for the inventories following non-bottleneck work centers.

Flexible Manufacturing Islands 98

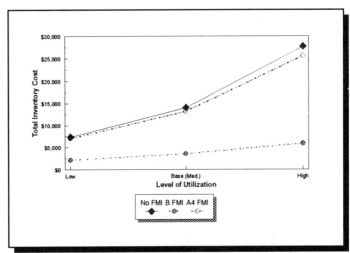

Figure 16. Comparison of Inventory Holding Cost: Capacity Utilization vs. FMI Position in Configuration III

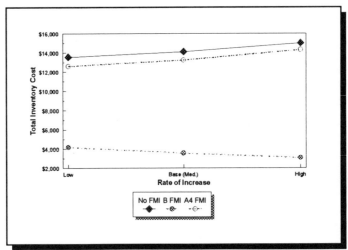

Figure 17. Inventory Holding Cost: Rate of Inventory Holding Cost Increase vs. FMI Position in Configuration III

Chapter 4: Results   99

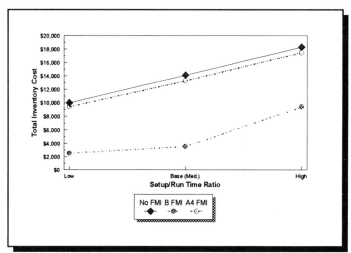

Figure 18. Comparison of Inventory Holding Cost: Setup/Run Time Ratio vs. FMI Position in Configuration III

Configuration II

System Overview

Configuration II is significantly different in structure from both configurations I and III. The bottleneck work center (B) is located in the center of the system with the two non-bottleneck work centers in front of and behind the bottleneck. Each non-bottleneck work center processes four items (versus two items per work center in configurations I and III). There are three operations in the routing of each items (versus two operations in configurations I and III). However, there are still two items per path through the system. Figure 19 shows the system schematic and item paths for configuration II.

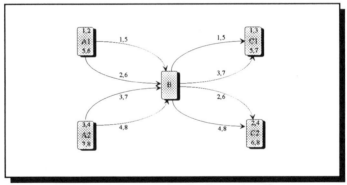

Figure 19. System Schematic for Configuration II

The base case parameters of the system under this configuration are presented in Table 41. The annual demand values for the items are the same as in configurations I and III, as are the setup and run time parameters for the bottleneck work center. The setup and run time parameters for the non-bottleneck work centers were set such that the

Chapter 4: Results   *101*

minimum cycle duration (C_{min}) values for these work centers are in the same range as they were for the non-bottleneck work centers in configurations I and III. The holding cost rate for each item at the bottleneck work center was set at the mid-point between the holding cost rates for the item at the front-end work center (A1 and A2) and the back-end work center (C1 and C2) in the item's routing.

Case 1: Conventional System

The conventional system involves no FMI and uses the system parameters presented in Table 41. In this case, again, the parameters were selected such that the bottleneck work center (B) determined the capacity of the system. The C_{min} for this work center is 48.11 days (as it was in the base cases for configurations I and III). This value establishes the cycle duration for the system, since it is the largest of the C_{min} for all work centers. The procedure for projecting the production schedule and inventory behavior following this work center was similar to other configurations. Table 42 shows the production cycle for work center B and the inventory behavior following the work center for item sequence A.

Flexible Manufacturing Islands

Table 41
System Parameters for Configuration II

Parameters	Items							
	1	2	3	4	5	6	7	8
System								
Demand	700	350	900	970	300	600	300	690
Work Center A1								
S	9.11	.....	.....	.....	9.22	.....	.....	.....
p	.05	.....	.....	.....	.25	.....	.....	.....
h	.02	.....	.....	.....	.06	.....	.....	.....
Work Center A2								
S	.....	7.02	.....	.....	.....	5.83	.....	.....
p	.....	.17	.....	.....	.....	.06	.....	.....
h	.....	.03	.....	.....	.....	.08	.....	.....
S	.....	.....	1.00	.....	.....	.....	1.58	.....
p	.....	.....	.10	.....	.....	.....	.35	.....
h	.....	.....	.04	.....	.....	.....	.09	.....
S	.....	.....	.....	9.26	.....	.....	.....	9.26
p	.....	.....	.....	.11	.....	.....	.....	.11
h	.....	.....	.....	.06	.....	.....	.....	.17
Work Center B								
S	4.50	3.5	2.75	2.25	3.00	5.00	2.25	.50
p	.03	.02	.04	.04	.04	.03	.05	.05
h	.03	.04	.08	.08	.11	.17	.17	.23

Legends:
  S = Setup times (days)
  p = Production times (per unit)
  h = Holding cost (dollar per unit per year)

Chapter 4: Results 103

Table 42
Production Schedule and Inventory Behavior Following Work Center B:
Configuration II, Conventional System, Base Case, Sequence A

Item	Lot Size (Units)	Start Setup (Day)	Start Prod. (Day)	End Prod. (Day)	Peak Amount (Units)	Start Cons.* (Day)	End Cons.* (Day)
1	93.56	0.00	4.50	7.31	93.56	7.31	10.11
2	46.78	7.31	10.81	11.74	46.78	46.78	46.78
3	120.29	11.74	14.49	19.30	120.29	19.30	19.30
4	129.64	19.30	21.55	26.74	129.64	26.74	26.74
5	40.10	26.74	29.74	31.34	40.10	31.34	31.34
6	80.19	31.34	36.34	38.75	80.19	38.75	38.75
7	40.10	38.75	41.00	43.00	40.10	43.00	43.00
8	92.22	43.00	43.50	48.11	82.22	48.11	48.11

* Start and end of consumption.

Note that the peak inventory amounts are equal to the lot sizes for the respective items and that production of item #8 end at day 48.11 which is the cycle duration for the system.

The relationships between work center B and the non-bottleneck work centers preceding it (A1 and A2) are similar to the same relationships in configuration III. Likewise, the relationships between work center B and the non-bottleneck work centers following it (C1 and C2) are similar to the relationships in configuration I. In other words, the starts of production for these items at work center B determine the ends of production for the same items at work centers A1 and A2. On the other hand, the starts of production for the items at work centers C1 and C2 are determined by the ends of production for the same items at work center B. The consumption rate for each item at work center B is equal to the production rate for the item at the back-end work center (C1 or C2) which process that item. For example, the consumption rate for item #1 at work center B is equal to 33.33 units per day which is equal to the production rate for this item at work center C1. To illustrate the link between non-bottleneck work

Flexible Manufacturing Islands 104

centers and work center B, production cycles for work centers A1 and C1 are discussed here. Table 43 shows the production cycle for work center A1, while Table 44 presents the same information for work center C1. Both tables are based on item sequence A. Note that item sequence A resulted in schedule conflict in both work centers A1 and C1. Tables 43 and 44 show the scheduling information before these conflicts were resolved. In work center A1, conflict is between items #1 and #2, in that the production of item #1 ends at day 4.50, while the setup for item #2 must start at day 3.56. This results in .94 days of conflict. The same type of conflict exists between items #3 and #5 at work center C1 (where the production of item #3 ends at day 27.72, while start of setup for item #5 should begin at day 26.24) resulting in 1.48 days of conflict). In fact, all of the five item sequences caused schedule conflict in the base case. The procedure for removing the conflicts for a given work center depended on whether the work center was preceding work center B or following it. If the work center with the conflict preceded work center B (i.e., work centers A1 or A2), the conflicts were resolved by pulling back the start of production for the first items involved in the conflict just enough to remove the conflict. For example, to resolve the conflict of .94 days between items #1 and #2, the production of item #1 was scheduled .94 days earlier. This shift resulted in carrying the inventory for this item following work center A1 for .94 days longer, which increased the inventory holding cost for this item.

Table 43
Production for Work Center A1: Configuration II
Conventional System, Base Case, Sequence A

Item	Start Setup (Days)	Start Prod.(Days)	End Prod.(Days)
1	-1.87	2.63	4.50
2	3.56	7.06	10.81
5	24.74	29.34	29.74
6	31.04	33.94	36.34

Chapter 4: Results 105

On the other hand, for conflicts occurring in the work centers following work center B (C1 and C2), the conflicts were resolved by delaying the start of the last item involved in the conflict. For example, to resolve the conflict between items #3 and #5 at work center C1, the start of production (and thus start of setup) for item #5 was delayed by 1.48 days. This change in schedule at work center C1 caused the inventory of item #5 following work center B to be held 1.48 days longer and increased the inventory holding cost for the item following work center B. Tables 45 and 46 show the production cycle and inventory behavior for work centers A1 and C1 for item sequence A after schedule conflicts at these work centers have been resolved.

Base Case

Since all five items sequences produced schedule conflict, clearly none of them was optimal. All five items sequences were evaluated to determine which resulted in the lowest inventory holding cost for the system. Table 47 shows the five items sequences and the system inventory holding cost for the sequences. Item sequence C resulted in the lowest system inventory cost and was considered the near-optimum sequence.

Table 44
Production Cycle for Work Center C1: Configuration II
Conventional System, Base Case, Sequence A

Item	Start Setup	Start Production	End Production
1	3.91	7.31	10.11
3	16.55	19.30	27.72
5	26.24	31.34	33.75
7	36.65	43.00	45.41

Table 45
Inventory Behavior Following Work Center A1:
Configuration II, Base Case, Conventional System, Sequence A

Item	Start Setup (Day)	Start Prod. (Day)	End Prod. (Day)	Peak Amt. (Units)	Start Cons* (Day)	End Cons* (Day)
1	-2.81	1.69	3.56	93.56	4.50	7.31
2	3.56	7.06	10.81	46.78	10.81	11.74
5	24.74	29.34	29.74	40.10	29.74	31.34
6	31.04	33.94	36.34	80.19	36.34	38.75

*Start and end of consumption.

Table 46
Inventory Behavior Following Work Center C1: Configuration II,
Conventional System, Base Case, Sequence A

Item	Start Setup (Day)	Start Prod. (Day)	End Prod. (Day)	Peak Amt. (Units)	Start Cons* (Day)	End Cons* (Day)
1	3.91	7.31	10.11	88.10	10.11	55.42
3	16.55	19.30	27.72	99.24	27.72	67.42
5	26.24	32.82	35.23	38.09	35.23	80.94
7	36.65	43.00	45.41	38.09	45.41	91.12

*Start and end of consumption.

Table 47
Inventory Holding Cost: Configuration II
Conventional System, Base Case, All Five Sequences

Item Sequence	Holding Cost
A	$15,286
B	15,620
C	15,253
D	15,673
E	15,292

Table 48
Inventory Holding Cost
Configuration II, Conventional System, Base Case, Sequence C

Work Center	Holding Cost
A1	$168
A2	1,379
B	2,326
C1	3,825
C2	7,555
Total	$15,253

Note that, for example, the cost for the best item sequence (sequence C) is only 2.7% lower than the cost for the worst sequence (sequence D) as shown in Table 48.

Sensitivity Analysis

The sensitivity analyses were carried out in the same manner as in configurations I and III. The parameters of the system were manipulated in the same way and the parameters used in the sensitivity analyses are shown in Appendix C. As was indicated in Chapter III, the sensitivity analysis on the rate of increase in holding cost rates was achieved by manipulating inventory holding cost parameters for all items at the first and last operation in their processing path. In manipulating the rate of increase in inventory holding cost rates, the holding cost parameters at work center B remained unchanged, while inventory holding costs per unit parameters at work centers A1 and A2 and at work centers C1 and C2 were manipulated. Item sequence C proved to be the near optimum sequence in all cases of the sensitivity analyses. Annual inventory holding cost for each work center and for the total system are presented in Table 49 for each of the sensitivity analyses. Table 50 compares the annual system holding costs for each sensitivity analysis to the base case.

Table 49
Total Inventory Holding Cost Under Sensitivity Analyses
Configuration II, Conventional System, Sequence C

Work Center	Capacity Utilization		Setup/Run Time		Rate of Holding Cost	
	Low	High	Low	High	Low	High
A1	$68	$442	$127	$235	$193	$143
A2	741	4,058	1,241	2,305	1,585	1,172
B	723	4,722	1,353	2,512	2,326	2,326
C1	2,017	7,158	2,677	4,972	3,445	4,204
C2	4,192	13,671	5,289	9,822	6,924	8,201
Total	$7,741	$30,051	$10,687	$19,846	$14,473	$16,046

Table 50
Comparison of Inventory Holding Costs Under the Base Case and the
Sensitivity Analyses: Configuration II, Conventional System; Sequence C

Level	Capacity Utilization	Setup/Run Time Ratio	Rate of Increase in Inventory Cost
High	$30,051 (1.97)	$19,846 (1.30)	$16,046 (1.05)
Base		$15,253 (1.00)	
Low	$7,741 (0.51)	$10,687 (0.70)	$14,473 (0.95)

[a] Figures in parentheses express the cost as a multiple of the base case cost.

Chapter 4: Results

Case 2: Work Center B is FMI

Conversion of work center B into FMI will change the status of this work center as a bottleneck work center. With work center B being FMI, the cycle duration of the system was determined by the 'bottleneck work center with the longest C_{min}. This work center was A2, with a minimum cycle duration of 27.87 days. Under this situation, work center A2 was the bottleneck work center and, therefore, established the basis for scheduling of the items processed by it (i.e., items #3, #4, #7, and #8). The most important impact of this change was that the meaning of "item sequence" as has been used so far changed. In establishing the production schedule for the bottleneck work centers, it was assumed that work center B would pass through all items at this work center with no delay. This assumption was made based on the fact that the production rates for all items at work center B are faster for their counterparts at the bottleneck work centers. Furthermore, the zero setup times for all items at work center B was an additional basis for the above assumption. Thus, "item sequence" had to be defined with respect to work center A2 and as far as this work center is concerned, the five item sequences selected collapses to two item sequences. These item sequences are 3-4-7-8 and 4-3-8-7 (these sequences correspond to the original item sequences A and C respectively). In other words, if viewed from a cyclical point, the other item sequences would be the same as the these two sequences. Thus, the production schedule for work centers A1, A2, B, C1, and C2 were determined for each of these item sequences. In general, the procedure of establishing production schedule for various work centers were as follows. For a given item sequence the production schedule of items in the sequence at work center A2 (i.e., #3, #4, #7, and #8) was first established. Then, based on the schedule in work center A2, production of these items at work centers C1 and C2 were determined. In establishing these links between work center A2 and these work centers, the production rates for various items were the primary factors. If the production rate for a given item at the front-end work center was faster than the production rate for the same item at the back-end work center, the start of the production times of these items at the two work center were linked together. For example, the production rate for item #8 at work center A2 is 16.67 units per day while this rate at work center C2 is 11.11 units per day. Therefore, in this case, the start of production of item #8 at work center C2 was linked to the start of production at work center A2. The reason for linking work centers A2 and C2 together in this fashion is that the linking the starting times of production will minimize the inventory build-up following work center A2. On the other hand, if the production rate at the back-end work center was faster for a given item, the end of production times were linked together. For example, the production rate for item #3 at work center A2 is 11.11 units per day while the production rate for this item at work center C1 is 14.29 units per day. Consequently, the end of production of item #3 at work center C1 was linked to the end of production for this item at work center A2. In this case, the logic of linking the end of production times at the two work centers was that if they both started processing at the same time, work center C1 would have been starved, because it is faster in processing item # 3 than work center A2. Once work centers C1 and C2 were scheduled for items #3, #4, #7, and #8, the free slot times in these work centers were used to schedule production of remaining items at the work centers (i.e., items #1, #2, #5, and #6). Then, based on the scheduling of these items at C1 and C2, the production schedule for work center A1 was established. In establishing these schedule, again the comparison of the production rates were the primary determinants, as previously explained.

Base Case

When item sequences A and C were evaluated, sequence A proved to result in lower holding cost for the system and, therefore, was the best sequence. Table 51 shows the inventory holding cost under the base case for configuration II when work center B is converted into FMI.

Chapter 4: Results

It is clear that converting work center B into an FMI has substantially reduced the inventory holding cost (from $15,253 in the conventional case). This reduction is caused by a shorter cycle duration, and therefore smaller lot sizes, as well as having no inventory following work center B.

Table 51
System Inventory Holding Cost
Configuration II, B is FMI, Base Case
Sequence A

Work Center	Inventory Cost
B	$0
A1	43
A2	223
C1	2,215
C2	4,376
Total	$6,857

Sensitivity Analysis

So far, in every case analyzed and discussed, every time sensitivity analysis was carried out, the cycle duration determined in the base case remained to be the system cycle duration. However, when work center B was converted into FMI and demand rates were manipulated to achieve low capacity utilization sensitivity analysis, the C_{min} of work center C1 became the largest, and therefore the cycle duration for the system. This change mandated evaluation of two new item sequences (as far as work center C1 is concerned) to determine which would result in lower holding cost. These item sequences are 1-3-5-7 and 1-7-5-3 (these sequences correspond to the original items sequences A and D respectively). When these item sequences were evaluated, the second sequence, i.e., item sequence D, caused no schedule conflict and was, therefore, the optimal sequence.

The parameters for sensitivity analyses were the same as the conventional case (see Appendix C), except that setup times for all items at work center B were now zero. Sensitivity analysis was carried out in the same manner as before using item sequence A in all case, except in low capacity utilization rate analysis. The annualized inventory holding cost for the various sensitivity analyses are shown in Table 52. Table 53 compares the annual inventory holding costs for each sensitivity analysis to the base case.

Table 52
Total Inventory Holding Cost Under Sensitivity Analyses
Configuration II, B is FMI, Sequence A and D

Work Center	Capacity Utilization		Setup/Run Time		Rate of Holding Cost	
	Low	High	Low	High	Low	High
A1	$17	$171	$30	$56	$49	$37
A2	90	726	156	290	257	190
B	0	0	0	0	0	0
C1	1,132	5,370	1,551	2,880	1,995	2,435
C2	2,616	10,582	3,063	5,689	4,010	4,750
Total	$4,044	$16,849	$4,800	$8,915	$6,311	$7,412

Table 53
Comparison of Inventory Holding Costs Under the Base Case and the Sensitivity Analyses
Configuration II, B is FMI; Sequence A and D

Level	Capacity Utilization	Setup/Run Time Ratio	Rate of Increase in Inventory Cost
High	$16,849 (2.45)	$8,915 (1.30)	$7,412 (1.08)
Base		$6857 (1.00)	
Low	$4,404 (0.64)	$4,800 (0.70)	$6,311 (0.92)

^a Figures in parentheses express the cost as a multiple of the base case cost.

Case 3: Work Center C2 is FMI

In this phase of analysis, work center C2 was converted into FMI. The reason for selecting this work center for conversion was that its inventory holding cost in the conventional system, base case, was higher than the other non-bottleneck work center, C1. Conversion of work center C2 into FMI did not affect the cycle duration for the system and it remained to be 48.11 as established by the C_{min} of work center B. Parameters of the system remained the same as the values shown in Table 39, except for setup time for items #2, #4, #6, and #8 at work center C2 were changed to zero. The production cycle and inventory behavior at work centers B, A1, A2, and C1 remained basically the same as the conventional system except that the consumption rates for the items #2, #4, #6, and #8 at work center B were now equal to their respective demand rates, rather than the production rates for these items at work center C2. This is due to the fact that work center C2, once being an FMI, allows unit flow of item to it. Also, the peak amounts for these items following work center B are not equal to their respective lot sizes (as was the case in configuration 1) but equal to the difference between production rates for these item at work center B and their demand rate (d) multiplied by length of time required to produce their lots. For example, the peak amount for item #2 following work center B is equal to 45.87 units which is equal to 49.03 (difference between production rate for item #2 at work center B, 50 units per day, and the demand rate for that items, .97 unit per day) multiplied by .94 (the length of time required to produce a lot of 46.78 units at the rate of 50 units per day at work center B). The same logic applies to items #4, #6, and #8.

The other distinguishing fact about configuration II when work center C2 is converted into FMI is that there was no inventory build-up following this work center for the items processed by it.

Again, the five item sequences were evaluate to determine which resulted in the lowest inventory holding cost for the

Chapter 4: Results 115

system. As in the conventional case, all five sequences resulted in some schedule conflict in work centers. Therefore, again, none of the sequences were optimal. However, when the inventory holding cost under the five item sequence were compared, item sequence E proved to be the lowest and thus the near-optimum sequence. Table 54 presents the five item sequences and the system inventory holding cost associated with each. Note, again, that the difference between the best item sequence and the worst item sequence with respect to the inventory cost is very small (as it was the case in the conventional case)

Table 54
System Inventory Holding Cost
Configuration II, C2 is FMI, Base Case
All Five Sequences

Item Sequence	Inventory Cost
A	$12,421
B	12,688
C	12,226
D	12,742
E	12,000

Flexible Manufacturing Islands 116

Table 55
System Inventory Holding Cost
Configuration II, C2 is FMI, Base Case
Sequence E

Work Center	Inventory Cost
A1	$168
A2	1,379
B	6,628
C1	3,825
C2	0
Total	$12,000

Table 55 shows the inventory holding cost for configuration II, base case, when work center C2 is FMI and under the near-optimum item sequence. Note that the inventory holding cost for work centers A1, A2, C1 are the same as the conventional system. This is, of course, because converting work center C2 into FMI did not affect their operations. Also, note the reduction of cost for work center B and work center C2. The reduction of inventory cost for work center B is explained by the fact that, as explained before, the peak amounts for items #2, #4, #6, and #8 were smaller than their respective lot sizes. However, this reduction in the inventory cost following work center B for the mentioned items is offset by the fact that the rates of consumption for these items, when C2 is FMI, are slower than when C2 was conventional (because they are now equal to the demand rates of these items). This made the consumption of inventory build-up following work center B more gradual and, thus, offset the decrease in the inventory holding cost. The inventory holding cost for work center C2 is zero since it is now an FMI. In effect, some of the holding cost for work center C2 in the conventional case is transferred to work center B when C2 becomes an FMI.

Chapter 4: Results 117

Sensitivity Analysis

Sensitivity analyses were carried out in the same manner as in conventional case. Similarly, the parameters of the system were manipulated in the same manner. The resulting parameters are presented in Appendix C. Item sequence E remained to be the near-optimum sequence under sensitivity analysis cases. The annualized inventory holding costs for the various sensitivity analyses are shown in Table 56. Table 57 compares the annual inventory holding costs for each sensitivity analysis to the base case.

Table 56
Total Inventory Holding Cost Under Sensitivity Analyses
Configuration II, C2 is FMI, Sequence E

Work Center	Capacity Utilization		Setup/Run Time		Rate of Holding Cost	
	Low	High	Low	High	Low	High
A1	$63	$410	$118	$218	$193	$143
A2	539	3466	965	1,792	1172	1,585
B	3,661	13,004	4,817	8,947	6,882	6,882
C1	2,117	6,938	2,677	4,972	4,204	3,445
C2	0	0	0	0	0	0
Total	$6,380	$2,3818	$8,577	$15,929	$12,105	$1,2401

Table 57
Comparison of Inventory Holding Costs Under the Base Case and the
Sensitivity Analyses
Configuration II, C2 is FMI; Sequence E

Level	Capacity Utilization	Setup/Run Time Ratio	Rate of Increase in Inventory Cost
High	$23,818 (1.98)	$15,929 (1.33)	$12,401 (1.03)
Base	$12000 (1.00)		
Low	$6,380 (0.53)	$8,577 (0.71)	$12,105 (1.01)

[a] Figures in parentheses express the cost as a multiple of the base case cost.

Chapter 4: Results 119

Case 4: Work Center A2 is FMI

In this phase of the analysis, work center A2 was converted into FMI. As prescribed by the methodology, the decision to convert this work center was based on the fact that the inventory holding cost for this work center ($1,379) was the higher than the other non-bottleneck work centers in the front-end of the system and. The parameters of the system remained as shown in Table 41, except that the setup times for all items processed by work center A2 were now equal to zero. Similar to configuration III when work center A4 was converted into FMI, work center A2 now allows unit-flow to and from it. This would have ordinarily eliminated inventory build-up following this work center. However, since work center B is still processing items #3, #4, #7, and #8 at a faster rate, some inventory has still to be build up following work center A2 in order to prevent starving work center B. The amount of this initial build up is determined by equation 5. These amounts were 66.83, 25.93, 28.30, and 15.37 for items #3, #4, #7, and #8 respectively. Thus, work center A2 continues to operate in batch mode even though it is an FMI now.

Base Case

Production of the required quantities in time to feed work center B caused some conflict in four of the five item sequences. Item sequences C, however, did not cause any conflict and was, therefore, optimal sequences. The inventory was built up and used up in the same manner as explained in configuration III when work center A4 was converted into FMI. The production cycle and inventory behavior for work centers B, A1, C1, and C2 were the same as the conventional case. Table 58 shows the inventory holding cost for configuration II when work center A2 is converted into FMI.

Note that the inventory costs for work centers A1, C1, C2, and B are the same as the conventional case. This is clearly explained by the fact that the conversion of work center A2 does not affect these work centers.

Table 58
Annual Inventory Holding Cost:
Configuration II, Base Case, A2 is FMI, Sequence C

Work Center	Inventory Cost
A1	$168
A2	303
B	2236
C1	3825
C2	7555
Total	$14,177

Sensitivity Analysis

The parameters used in the sensitivity analysis are the same as for the conventional system (see Appendix C), except that the setup times for all items at work center A2 are zero. The annualized inventory holding costs for the various sensitivity analyses are shown in Table 59. Table 60 compares the annual inventory holding costs for each sensitivity analysis to the base case.

Chapter 4: Results

Table 59
Total Inventory Holding Cost Under Sensitivity Analyses
Configuration II, A2 is FMI, Sequence C

Work Center	Capacity Utilization		Setup/Run Time		Rate of Holding Cost	
	Low	High	Low	High	Low	High
A1	$723	$4722	$1353	$2512	$2326	$2326
A2	68	442	127	235	208	154
B	114	793	212	394	348	258
C1	2,017	7,158	2,677	4,972	3,766	4,204
C2	4,192	13,671	5,289	9,822	7,196	8,201
Total	$7,114	$26,786	$9,658	$17,935	$13,844	$15,143

Table 60
Comparison of Inventory Holding Costs Under the Base Case and the
Sensitivity Analyses
Configuration II, A2 is FMI; Sequence C

Level	Capacity Utilization	Setup/Run Time Ratio	Rate of Increase in Inventory Cost
High	$26,786 (1.89)	$17,935 (1.27)	$15,143 (1.07)
Base		$14,177 1.00	
Low	$7,114 (0.50)	$9,658 (0.68)	$13,844 (0.98)

ª Figures in parentheses express the cost as a multiple of the base case cost.

Flexible Manufacturing Islands   122

The results of the analysis for configuration II are summarized in Table 61. The annual inventory holding cost is shown for the conventional system and for both positions of FMI under the base case and under the various sensitivity analyses. The figures in parentheses express the cost for that case as a multiple (or percentage) of the cost for the conventional ("NO FMI") case in that same column. Figures 20, 21, and 22 graphically show the comparison of the holding costs presented in Table 61.

Table 61
Total Inventory Holding Cost Under Various Cases and FMI Positions: Configuration II

FMI Position	Base Case	Capacity Utilization		Setup/Run Time		Rate of Holding Cost	
		Low	High	Low	High	Low	High
No FMI	15253 (1.00)	7741 (1.00)	30051 (1.00)	10687 (1.00)	19846 (1.00)	14473 (1.00)	16046 (1.00)
A2 FMI	14177 (0.93)	7114 (0.92)	26786 (0.98)	9658 (0.90)	17935 (0.90)	1384 (0.89)	15143 (0.95)
B FMI	6857 (0.45)	4044 (0.52)	16849 (0.56)	4800 (0.45)	8915 (0.45)	6311 (0.41)	7412 (0.46)
C2 FMI	12000 (0.79)	6380 (0.82)	23818 (0.79)	8577 (0.80)	15929 (0.80)	12105 (0.78)	12401 (0.78)

Chapter 4: Results   *123*

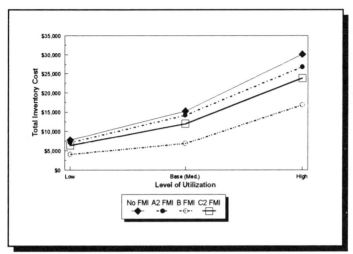

Figure 20. Comparison of Inventory Holding Cost: Capacity Utilization vs. FMI Position, Configuration II

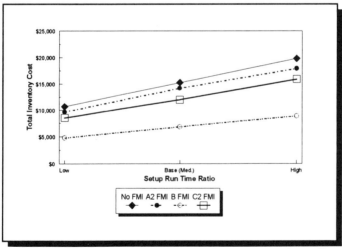

Figure 21. Comparison of Inventory Holding Cost: Setup/Run Time Ratio vs. FMI Position, Configuration II

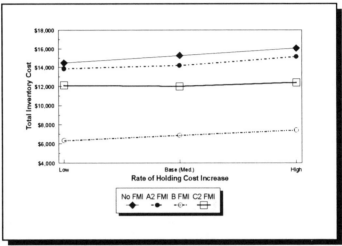

Figure 22. Inventory Holding Cost: Rate of Inventory Holding Cost Increase vs. FMI Position, Configuration II

Chapter 4: Results

The following observations can be made on the results for configuration II as presented in Table 61 and in Figures 21, 22, and 23.:

- Converting a work center into an FMI reduces inventory holding cost for the system, regardless of the position of the work center in the system.
- The reduction in the system inventory holding cost is greatest when the work center converted into an FMI is the bottleneck work center (B).
- When a non-bottleneck work center is converted into an FMI, the reduction in system inventory holding cost is greater when the FMI is at the back end of the system rather than the front end.
- Inventory holding cost is very sensitive to capacity utilization and the relationship is non-linear.
- The relationship between inventory holding cost and the ratio of setup time to run time is approximately linear within the range of sensitivity analysis conducted. Tables 50, 53, and 60 show that a 30 percent increase (decrease) in the setup to run time ratio resulted in an approximate 30 percent (decrease) in inventory holding cost for the system in the conventional system and for each position of the FMI.
- Inventory holding cost is relatively insensitive to changes in the rate of increase in holding cost.

In Brief

In this chapter, the results of the research were presented. The presentation of the results was made in three phases, each corresponding to a system configuration. For each configuration, the general characteristics of the system were described. The results for that system configuration under the conventional and the mixed convention/FMI system structures were presented, discussed, and compared. Under each configuration, the results of the sensitivity analyses were conducted and the results were compared to the based case. The results and comparisons were presented in both tabular

Flexible Manufacturing Islands 126

and graphical forms. Table 62 shows the complete results of the analyses for all configurations. Table 63 shows the percentage reduction in inventory holding cost under various configurations and FMI positions in each configuration. the percentages in this table show how much the system inventory holding cost was reduced when a work center was converted into an FMI as compared to the conventional system (base case) in a given configuration.

Table 62
System Total Annual Inventory Holding Cost
Under Various Configuration and FMI Positions

FMI Position	Base Case	Capacity Utilization		Setup/Run Time		Rate of Holding Cost	
		Low	High	Low	High	Low	High
CONFIGURATION I							
No FMI	$14,056	$27,408	$7,222	$18,256	$98,830	$15,355	$13,751
B FMI	8,393	15,393	4,702	11,300	5,667	9,614	8,414
C4 FMI	10,740	20,138	5,745	13,947	7,510	1,105	11,425
CONFIGURATION II							
No FMI	$15,253	$30,051	$7,741	$19,846	$10,687	$15,942	$15,535
A2 FMI	14,177	26,786	7,114	17,935	9,658	15,143	13,844
B FMI	6,856	16,818	4,044	8,914	4,800	7,411	6,310
C2 FMI	12,000	23,818	6,380	15,929	8,577	12,401	12,105
CONFIGURATION III							
No FMI	$14,288	$27,707	$7,398	$18,316	$10,001	$15,008	$13,569
A4 FMI	13,182	25,007	6,984	17,136	9,227	14,165	12,200
B FMI	3,569	5,889	2,135	4,641	2,498	3,092	4,183

Table 63
Percentage Reduction in Inventory Holding Cost of the System
Under Various Configuration and FMI Position

FMI Position	Configuration I	Configuration II	Configuration III
A_i	N/A	7.05%	4.58%
B	38.70%	55.04%	74.67%
C_i	23.59%	21.33%	N/A

CHAPTER V

CONCLUSIONS AND RECOMMENDATIONS FOR FUTURE RESEARCH

In recent years there has been much interest in automating manufacturing operations. Benefits claimed for factory automation are believed to create competitivity advantage for those firms that employsuch technologies effectively. However, research indicates that incremental approaches to technological adaptation are more prevelant than a completed conversion of a conventional manufacturing system into an automated factory. The purpose of this study was to investigate various issues related to the incremental conversions of work centers in a manufacturing system into Flexible Manufacturing Islands (FMI). Research and literature relevant to the issueof this study were reviewed in Chapter II and the methodology employed in this study was explained in Chapter III by using a simplified example. Three specific systems configurations were investigated, and the results were presented in Chapter IV. Brief observations were included at the end of the discussion for each system configuration. In this chapter, the overall conclusions of the study are presented. In addition, recommendations as to potential extensions of this research are discussed.

Conclusions

Based on the results presented in Chapter IV, the following conclusions are drawn. These conclusions apply only within the scope, limitations, and assumptions of this study.

1. For all of the system configurations examined in this study, conversions of a work center in to an FMI leads to a reduction in inventory holding costs for the system as a whole. This is true regardless of the positon in the system of the work center that is converted into an FMI. However, converting a given work center into an FMI may increase the inventory holding cost for other work centers under specific configurations. For example, in configuration III, converting the bottleneck work center (B) into an FMI resulted in increasing the inventoryholding

Chapter 5: Conclusions 129

cost for all of the non-bottleneck work centers. However, inventory holding costs for the system as a whole decreased because the reduction in the holding cost for work center B more than offset the increase in the costs for the non-bottleneck work centers. The explanations for the reduction in inventory holding cost when a work center is converted into an FMI are as follows:

   a. Concerting the bottleneck work center into and FMI results in a shorter production cycle duration for the system, which in turn results in smaller lot sizes. Therefore, inventory peaks are reduced/
   b. When a work center is converted into an FMI, unit-flow of material to and from the work center is possible, and this reduces work-in-process inventory between successive work centers. Even when some inventory buildup following such work centers is necessary (for example, when work center A4 in configuration III is converted into an FMI), the peak inventory amounts following such work centers are smaller than the respective lot sizes for the items.
   c. When the work center converted into an FMI is positioned at the back end of the systems configuration, it completely eliminates the inventory of finished goods for the items processed by the work center.

2. Converting a bottleneck work center (work Center B in all configurations) into an FMI invariably results in a higher reduction in inventory holding cost than is realized by the conversion of any non-bottleneck work center. This result is clearly due to the effects explained in a 1a above.

3. When a non-bottleneck work center is converted into an FMI, it has more impact in reducing inventory holding cost if the converted work center is positioned at the back end of the system. This is because of the facts that (a) inventory holding cost per unit increases as the items are processed by successive work centers and (b) the conversion

of a back and work center completely eliminates the inventory of finished goods following that work center.

4. Regardless of the system configuration or positon of the FMI, the system inventory holding cost seems to react similarly to changes in capacity utilization rate and the changes in the ratio of setup time to run time. More specifically, the following conclusions can be drawn regarding the relationship between inventory holding cost and the capacity utilization rate:

   a. Inventory holding cost is very sensitive to changes in capacity utilization rate.
   b. The relationship is positive and non-linear. When the capacity utilization rate approaches 100%, the invntory holding cost will clearly appraoch infinity.

   Similarly, the following observations can be made regarding the relationship between setup time to run time ratio and inventory holding cost:

   a. Inventory holding cost is less snesitive tothe setup to run time ration than to capacity utilization rate.
   b. The relationship is positive and approximately linear within the ranges of sensitivity analysis conducted in the study.

5. The following conclusions cna be drawn concerning the relationship between the total system inventory cost and the rate of increase in inventory holding cost rates:

   a. Inventory holding cost if relatively insensitive to changes in the rate of increase in holding cost rates.
   b. The relationship between rate of increase in holding costs rates and system inventoryholding cost is approximately linear. However, relationship may be positive or negative depending upon specific case and configuration.

Suggestions for Further Research

Several issues which are not addressed by this study may warrant further research. Among the issues related to this study which could be fertile ground for further research are:

- ☞ Investigating the incremental impact of converting a second, a third, or more work centers into FMIs.
- ☞ Investigating more complex system configurations.
- ☞ Investigating the situation when setup times are dependent on item sequence.
- ☞ Examining the sensitivity of the results to such problems as scrap rates, machine break-downs, and quality problems.
- ☞ Investigating the relationship between this study and real-world cases of incremental installation of FMS.
- ☞ Integration of this line of research into other research dealing with such issues as scheduling of AGVs, loading of work centers, and tooling problems.

APPENDIX

Parameters of the system under sensitivity analysis

Table 64
Annual Demand Under Sensitivity Analysis: Configuration I

Item	Annual Demand	
	High	Low
1	910	490
2	455	245
3	1170	630
4	1261	679
5	390	210
6	780	420
7	390	210
8	879	483

Table 65
Setup Times Under Sensitivity Analysis: Configuration 1

	Level of Sensitivity Analysis										
	LOW LEVEL					HIGH LEVEL				
Item	WC B	WC C1	WC C2	WC C3	WC C4	WC B	WC C1	WC C3	WC C3	WC-C4
1	3.15	6.38	...	...	...	5.85	11.84	...	...	...
2	2.45	...	4.91	...	...	4.55	...	9.13	...	...
3	1.92	...	...	0.70	...	3.58	...	...	1.30	...
4	1.57	...	...	...	6.48	2.93	...	...	...	12.04
5	2.10	6.45	...	...	...	3.90	11.99	...	...	...
6	3.50	...	4.08	...	...	6.50	...	7.58	...	...
7.	1.57	...	...	1.11	...	3.93	...	...	2.05	...
8	0.35	...	...	...	6.37	0.65	...	...	...	12.51

Table 66
Inventory Holding Costs/Unit Under Sensitivity Analysis

Item	Level of Sensitivity Analysis									
	LOW LEVEL					HIGH LEVEL				
	WC B	WC C1	WC C2	WC C3	WC C4	WC B	WV-C1	WC C2	WC C3	WC-C4
1	.026	.026	...	...	...	.020	.034	...	...	...
2	.094	...	.076	...	...	.025	...	.085	...	...
3	.046	...	...	.074	...	.034	...	...	.086	...
4	.069	...	...	...	.071	.051	...	...	...	.089
5	.069	.101	...	...	...	.051	.119	...	...	...
6	.092	...	.098	...	...	.068	...	.122	...	...
7	.103	...	...	.157	...	.076	...	...	.183	...
8	.196	...	...	...	.204	.145	...	...	...	.256

Table 67
Annual Demand Under Sensitivity Analysis Configuration II

Item	Annual Demand	
	High	Low
1	910	490
2	455	245
3	1170	630
4	1261	679
5	390	210
6	780	420
7	390	210
8	897	483

Table 68
Setup Times Under Sensitivity Analysis: Configuration II

	Level of Sensitivity Analysis									
	LOW LEVEL					HIGH LEVEL				
Item	WC-B	WC-A1	WC-A2	WC-C1	WC-C2	WC-B	WC-A1	WC-A2	WC-C1	WC-C2
1	3.15	3.15	...	2.38	...	5.85	5.85	...	4.42	...
2	2.45	2.45	...	...	0.94	4.55	4.55	...	...	1.76
3	1.92	...	0.35	1.92	...	3.58	...	0.65	3.58	...
4	1.57	...	3.22	...	3.29	2.93	...	4.98	...	6.11
5	2.10	3.22	...	3.57	...	3.90	5.98	...	6.63	...
6	3.50	2.03	...	...	1.82	6.50	3.77	...	...	3.38
7	1.57	...	0.55	4.44	...	2.93	...	1.03	8.25	...
8	0.35	...	3.36	...	2.77	0.65	...	6.24	...	5.14

Appendix   135

Table 69
Inventory Holding Costs/Unit Under Sensitivity Analysis: Configuration II

Item	Level of Sensitivity Analysis							
	LOW LEVEL				HIGH LEVEL			
	WC-B	WC-A1	WC-A2	WC-C1	WC-C2	WC-B	WC-A1	WC-A2	WC-C1	WC-C2
1	.025	.023	...	.027	...	.025	.017	...	.033	...
2	.035	.034	...	...	.036	.035	.025	...	...	.044
3	.070	...	.069	.071	...	.070	...	.051	.089	...
4	.070	...	.046	...	.074	.070	...	.034	...	.086
5	.095	.092	...	.098	...	.095	.068	...	.122	...
6	.092	.069	...	...	.161	.092	.051	...	...	.179
7	.102	...	.098	.157	...	.102	...	.072	.183	...
8	.200	...	.196	...	.205	.200	...	.146	...	.256

Table 70
Annual Demand Under Sensitivity Analysis: Configuration III

Item	Annual Demand	
	High	Low
1	910	490
2	455	245
3	1170	630
4	1261	679
5	390	210
6	780	420
7	390	210
8	897	483

Table 71
Setup Times Under Sensitivity Analysis: Configuration III

Item	Level of Sensitivity Analysis									
	LOW LEVEL					HIGH LEVEL				
	WC B	WC A1	WC A2	WC A3	WC A4	WC B	WC A1	WC A2	WC A3	WC A4
1	3.15	6.38	.....	.....	.....	5.85	11.84	.....	.....	.....
2	2.45	.....	4.91	.....	.....	4.55	.....	9.13	.....	.....
3	1.92	.....	.....	0.70	.....	3.58	.....	.....	1.30	.....
4	1.57	.....	.....	.....	4.08	2.93	.....	.....	.....	4.08
5	2.10	6.45	.....	.....	.....	3.90	4.08	.....	.....	.....
6	3.50	.....	4.08	.....	.....	6.50	.....	4.08	.....	.....
7	1.57	.....	.....	4.08	.....	2.93	.....	.....	4.08	.....
8	0.35	.....	.....	.....	4.08	0.65	.....	.....	.....	4.08

Table 72
Inventory Holding Costs/Unit Under Sensitivity Analysis: Configuration III

Item	Level of Sensitivity Analysis									
	LOW LEVEL					HIGH LEVEL				
	WC-B	WC-A1	WC-A2	WC-A3	WC-A4	WC-B	WC-A1	WC-A2	WC-C1	WC-C2
1	.027	.026	....	....	....	.034	.020	....	....	....
2	.036	....	.034	....	....	.045	....	.025	....	....
3	.074	....	....	.046	....	.086	....	....	.034	....
4	.071	....	....	....	.069	.089	....	....	....	.051
5	.101	.069	....	....	....	.119	.051	....	....	....
6	.158	....	.092	....	....	.182	....	.068	....	....
7	.157	....	....	.103	....	.183	....	....	.076	....
8	.205	....	....	....	.196	.256	....	....	....	.145

REFERENCES

1. Abernathy, W.J., K.B. Clark, and A.M. Kanton, "The New Industrial Competition," *Harvard Business Review*, 59 (September/October 1981):79

2. Afentakis, P., "A Model for Layout Design in FMS," in A. Kusiak (ed.) *Flexible Manufacturing Systems: Methods and Studies* (New York: Elsevier Science Publishers, 1986), pp. 127-139

3. Afentakis, P., "Maximum Throughput in Flexible Manufacturing Systems," in K.E. Stecke (ed.) *Proceedings of the Second ORSA/TIMS Conference on Flexible Manufacturing System: Operation Research Models and Applications* (New York: Elsevier Science Publishers, 1986), pp. 333-343.

4. Afentakis, P.B., B. Gavish, and U.S. Karmarkar, "Exact Solution to the Lot-Sizing Problem in Multistage Assembly Systems," *Management Science*, 30 (1984):222-239.

5. Afentakis, P. R.A. Millen and M.M. Solomon, "Layout Design for Flexible Manufacturing Systems: Models and Strategies," In K.E. Stecke (ed.) *Proceedings of the Second ORSA/TIMS Conference on Flexible Manufacturing Systems: Operations Research Models and Applications* (New York: Elsevier Science Publishers, 1986), pp. 221-228

6. Askin, R.G., "A Procedure for Production Lot Sizing with Probabilistic Dynamic Demand," *AIIE Transactions*, 13 (1981): 132-137.

7. Blynsky, Gene, "The Rate to the Automatic Factory," *Fortune*, 21 February 1983, p. 54.

8. Brown, E., "Flexible Work Stations Offer Improved, Cost-Effective Alternative for Factory," *Industrial Engineering*, June 1985, pp. 50-59

9. Buffa, E.S., *Meeting the Competitive Challenge: Manufacturing Strategies for U.S. Companies* (Illinois: Richard Irwin, 1984), p. 83

10. Burstein, M.C., "Finding the Economical Mix of Rigid and Flexible Automation for Manufacturing Systems," in K.E. Stecke (ed.), *Proceedings of the Second ORSA/TIMS Conference on Flexible Manufacturing Systems: Operations Research Models and Applications* (New York: Elsevier Science Publishers, 1986), pp. 43-54

11. Buzacott, J.A. and J.G. Shanthikumar, "Models for Understanding Flexible Manufacturing System," *AIIE Transactions*, 12 (1980): 339-350

12. Bylinsk, G." The Race to the Automatic Factory," *Fortune*, February 21, 1983, p. 54

13. Chakravaty, A.K. and A. Schtub, "Production Planning with Flexibilities in Capacity," in K.E. Stecke (ed.) *Proceedings of the Second ORSA/TIMS Conference on Flexible Manufacturing System: Operation Research Models and Applications* (New York: Elsevier Science Publishers, 1986), pp. 333-343.

14. Chang, Y.L. and R.S. Sullivan, "Lot-Sizing in a Flexible Assembly System," in K.E. Stecke (ed.), *Proceedings of the Second ORSA/TIMS Conference on Flexible Manufacturing Systems: Operation Research Models and Applications* (New York: Elsevier Science Publishers, 1986), pp. 359-368

15. Church, J., "Simulation Aspects of Flexible Manufacturing Design and Analysis," *1982 Annual Industrial Engineering Conference Proceedings*, Cincinnati, Ohio (1982), p. 427.

16. Church, J., "Scheduling Batch Assembly," *SME Technical Paper, No. AD74-227*, 1974.

17. Co, H.C., Jan, T.J., and Chen, S.K., "Sequencing in Flexible Manufacturing Systems and Other Short Queue-Length Systems," *Journal of Manufacturing Systems*, 7 (1988): 1-8.

18. Costa, A. and M. Garetti, "Design of a Control System for a Flexible Manufacturing Cell," *Journal of Manufacturing Systems*, 4 (1985):65-84

19. Cudworth, E.F., "Pratt and Whitney's $200 Million Factory Showcase," *Industrial Engineering*, October 1984, pp. 90-97

20. Cutosky, M.R., P.S. Fussel, and R. Milligan, "The Design of a Flexible Manufacturing Cell for Small Batch Production," *Journal of Manufacturing Systems*, 3 (1984):39-59

21. Dannenbring, D.G., "An Evaluation of Flow Shop Sequencing Heuristics," *Management Science*, 23 (1977):1174-1182.

22. Dar-El, E. and S. Cucuy, "Optimal Mixed-Model Sequencing for Balanced Assembly Lines, " *OMEGA: The International Journal of Management Science*, 5 (1977): 333-342.

23. Delporte, C.M. and L. J. Thomas, "Lot Sizing and Sequencing for N Products on One Facility," *Management Science*, 23 (1977): 1070-1079.

24. Ferrario, J.D., "Integration of Flexible Systems with the Factory Environment and the Work Force," *1983 Fall Industrial Engineering Conference*, Toronto, Canada, November 1983, p.388

25. Flynn, B.B. and F.R. Jacobs, "An Experimental Comparison of Cellular (Group Technology) Layout with Process Layout," *Decision Sciences*, 18 (1987):562-581

26. Galvin, T., "Economic Lot-Scheduling Problem with Sequence- Dependent Setup Costs," *Production and Inventory Management*, 28 (1987): 96-104.

27. Greshwin, Stanley B., "Stochastic Scheduling and Setups in Flexible Manufacturing Systems," in K.E. Stecke (ed.) *Proceedings of the Second ORSA/TIMS Conference on Flexible Manufacturing System: Operation Research Models and Applications* (New York: Elsevier Science Publishers, 1986), pp. 431-442.

28. Gomma, H., "Computer Integrated Manufacturing Architecture of FMS," *SME Technical Paper, No MS86-159*, 1986

29. Graves, S.C., "A Review of Production Scheduling, " *Operations Research*, 29 (1981):646-675.

30. Graves, S.C., "The Multi-Product Production Cycling Problem," *AIIE Transactions*, 12 (1983):233-240.

31. Hall, D. and K.E. Stecke, "Design Problems of Flexible Manufacturing Assembly Systems," in K.E. Stecke (ed.), *Proceedings of the Second ORSA/TIMS Conference on Flexible Manufacturing Systems: Operations Research Models and Applications* (New York: Elsevier Science Publications, 1986), pp. 145-156

32. Hartley, J. *FMS at Work* (United Kingdom: IFS Publications, 1984), Chapter 10

33. Hartley, J. *FMS at Work* (United Kingdom: IFS Publications, 1984), Chapter 10

34. Hartley, J. *FMS at Work* (United Kingdom: IFS Publications, 1984), Chapter 10

35. Harvey, R.E., "Factory 2000," *Iron Age*, June 4, 1984, pp. 72-76

36. Hayes, R.H. and S.C. Wheelwrigth, *Restoring Our Competitive Edge Through Manufacturing* (New York: John Wiley, 1984), p. 192

37. Hildebrant, R.R., "Scheduling Flexible Machining System Using Mean Value Analysis," *Proceedings of the 19th. IEEE Conference* on Decision Control, Albuquerque, NM (December 1980), pp. 701-706.

38. Hill, M.R., "FMS Management: The Scope for Future Research," *International Journal of Operations and Production Management*, 5 (1985): 6.

39. Hughes, T. and D. Hegland, "Flexible Manufacturing: The Way to the Winners Circle," *Production Engineering*, 30 (September 1983):55

40. *Industrial Engineering*, issues beginning with January 1984

41. Jain, S. and W.J. Foley, "Basis for Development of a Generic FMS Simulator, " in K.E. Stecke (ed.), *Proceedings of the Second ORSA/TIMS Conference on Flexible Manufacturing Systems: Operations Research Models and Applications* (Amsterdam: Elsevier Science Publishers, 1986), p. 394.

42. Kaplan, R.S., "Yesterday's Accounting Undermines Production," *Harvard Business Review*, 61 (1984):95-101.

43. Karmarkar, U.S., "Lot-Sizes, Lead Times, and In-Process Inventories," *Management Science*, 33 (1987)3: 409-418.

44. Karmarkar, U. and L. Scharge, "The Deterministic Dynamic Product Cycling Problem," *Operations Research*, 31 (1985):326-345.

45. Karmarkar, U., S. Kekre, and S. Kekre, "Multi-Item Lot Sizing and Lead Times," *Working Paper No. QM8325*, Graduate School of Management, University of Rochester (1983).

46. Karmarkar, U.S., S. Kekre, and S. Kekre, "Lot Sizing in Multi-Item, Multi-Machine Job Shops," *AIIE Transactions*, 17 (1985):290-298.

47. Kiran, A.S., "The System Setup in FMS: Concepts and Formulation," in K.E. Stecke, (ed.) *Proceedings of the Second ORSA/TIMS Conference on Flexible Manufacturing Systems: Operations Research Models and Applications* (New York: Elsevier Science Publishers, 1986), pp. 321-332

48. Klahorst, T.H. "Flexible Manufacturing Systems: Combining Elements to Lower Costs, Add Flexibility," *Industrial Engineering* November 1981, pp. 112-117

49. Kusiak, A., "Scheduling Flexible Machining and Assembly Systems," in K.E. Stecke (ed.) *Proceedings of the Second ORSA/TIMS Conference on Flexible Manufacturing System: Operation Research Models and Applications* (New York: Elsevier Science Publishers, 1986), pp. 521-532.

50. Leachman, R.C. and A. Gascon, "A Heuristic Policy for Multi-Item, Single-Machine Production Systems With Time Varying Stochastic Demand," *Management Science*, 34 (1988):1174-1182.

51. Litteral, L.A.,"Single Machine Sequencing Techniques with Both Tardiness and Flow-time Criteria," *Production and Inventory Management*, 28 (1987):59-63.

52. Mabert, V.A. and P.A. Pinto, "Product Batch Sizes in a Repetitive Flexible Assembly System," *Production and Inventory Management*, 28 (1987):34-37.

53. Maimon, O.Z., "Real-Time Operational Control of Flexible Manufacturing Systems," *Journal of Manufacturing Systems*, 6 (1987):125-136

54. Maimon, O.Z., and Y.F. Choong, "Dynamic Routing in Reentrant FMS," in K.E. Stecke (ed.) *Proceedings of the Second ORSA/TIMS Conference on Flexible Manufacturing System: Operation Research Models and Applications* (New York: Elsevier Science Publishers, 1986), pp. 467-475.

55. Malone, R., "FMS Thrives," *Managing Automation*, August 1987, p.27

56. *Managing Automation*, August 1987, pp. 27-65

57. Mize, J.H. and D.J. Seifiet, "CIM- A Global View of the Factory," *1985 Annual International Industrial Engineering Conference Proceedings* (Institute of Industrial Engineers, 1985), p. 173

58. Nawaz, M., E. Enscore, Jr., and I. Ham, "A Heuristic Algorithm for the m-Machine, n-Job Flow-Shop Sequencing Problem," *OMEGA: The International Journal of Management Science*, 11 (1983):91-95.

59. Office of Technology Management, *Computerized Manufacturing Automation: Employment, Education, and the Work Place* (Washington, D.C.: OTA, 1984)

60. Piszczalski, M., "Strategies for Spending Millions," *Managing Automation*, August 1987, p. 32

61 Ranky, Paul, G., *The Design and Operation of FMS* (UK: IFS Publication, 1983).

62. Salomon, D.P. and J.E. Biegel, "Assessing Economic Attractiveness of FMS Applications in Small Batch Manufacturing," *Industrial Engineering*, January 1984, pp. 88-96

63. Shannon, R.E. and D.T. Phillips, "Comparison of Modelling Languages for Simulation of Automated Manufacturing Systems," *Autofact 5 Conference Proceedings*, Detroit, Michigan (1983)

64. Shin, H. and W.E. Wilhelm, "Hierarchial Planning and Control of Alternate Operations in a Flexible Manufacturing System," *1983 Fall Industrial Engineering Conference Proceedings*, Toronto, Canada, November 1983, pp. 369-374.

65. Smith-Daniel, V.I. L.P. Ritzman, "A Model for Lot Sizing and Sequencing in Process Industries," *International Journal of Production Research*, 26 (1988): 647-674.

66. Spur, G., G. Selinger, and B. Viehweger, "Cell Concepts for Flexible Automated Manufacturing," *Journal of Manufacturing Systems*, 5 (1984):7

67. Stecke, K.E. and I.Y. Kim, "A Flexible Approach to Implementing the Short Term FMS Planning Function," in K.E. Stecke (ed.) *Proceedings of the Second ORSA/TIMS Conference on Flexible Manufacturing System: Operation Research Models and Applications* (New York: Elsevier Science Publishers, 1986), pp. 283-295.

68. Stecke, K.E. and T.L. Morrion, "The Optimality of Balancing Workloads in Certain Types of Flexible Manufacturing Systems," *European Journal of Operational Research*, 20 (1985):68-82

69. Suri, R. and C.K. Whitney, "Decision Support Requirements in Flexible Manufacturing," *Journal of Manufacturing Systems*, 3 (1984):61

70. Suri, R. and C.K. Whitney, "Decision Support Requirements in Flexible Manufacturing," *Journal of Manufacturing Systems*, 3 (1984): 61

71. Suri, R. and R.R. Hildebrant, "Modelling Flexible Manufacturing Systems, Using Mean-Value Analysis," *Journal of Manufacturing Systems*, 3 (1984):27

72. Szendrovits, A.Z., "Manufacturing Cycle Time Determination for a Multi-Stage Economic Production Quantity Model," *Management Science*, 22 (1975):298-308.

73. Taha, H.A. and R.W. Skeith, "The Economic Lot-Sizes in Multi- Stage Production Systems," *AIIE Transactions*, 2 (1970): 157-162.

74. Talaysum, A.T., Z. Hassan, and J.P. Goldhar, "Scale vs. Scope Considerations in the CIM/FMS Factory," in A. Kusiak (ed.) *Flexible Manufacturing Systems: Methods and Studies* (New York, Elsevier Science Publishers, 1986), pp. 45-54

75. Tersine, R.J., *Material Management and Inventory Systems* (New York: North Holland Publishing Co., 1978), pp. 143-147.

76. Thomopolous, Nick T., *Mixed Model Line Balancing with Smoothed Station Assignment* (New York: North Holland Publishing Co., 1978), pp. 143-147.

77. Vachajitpan, P., "Job Sequencing with Continuous Machine Operation," *Computers and Industrial Engineering*, 6(1982): 255-59.

78. Vaithianathan, R., "Scheduling in Flexible Manufacturing Systems," *1982 Fall Industrial Engineering Conference*, Cincinnati, Ohio, 1982.

79. Valcada, A. and M. Mastratta, "Comparison of Different Computerized Design Tools for Flexible Manufacturing," *Proceedings of the 3rd. International Conference on Flexible Manufacturing Systems*, Boeblingen, West Germany, 1984, pp. 387-395

80. Venkitaswamy, R., "FMS Around the World," *SME Technical Papers, No. MS86-163*, 1986, p. 3.

81. Wagner, H.M. and T.M. Within, "Dynamic Version of the Economic Lot-Size Model," *Management Science*, 5 (1958):89-96

82. Wemmerlov, U. and N.L. Hyer, "Research Issues in Cellular Manufacturing," *International Journal of Production Research*, 25 (1987):418-422

83. Wester L. and M.D. Kilbridge, "The Assembly Line Mixed Model Sequencing Problem," in T.O. Prenting and N.T. Thomopolous (eds.), *Humanism and Technology in Assembly Line Systems* (New Jersey, Hyden Book Co., 1974), pp. 210-222.

84. Wilhelm, W.E. and S.C. Sarin, "Models for the Design of Flexible Manufacturing Systems," *Proceedings of the Annual AIIE Conference*, Louisville, KY, 1983, pp. 564-574

85. Wilhelm, W.E. and S.C. Sarin, "Models for the Design of Flexible Manufacturing Systems," *1983 Annual Industrial Engineering Conference Proceedings* Louisville, KY (1983), pp. 564-574.

86. Wortman, D.B., "Simulation and Flexible Manufacturing Systems: Partners in Productivity," *SME Technical Papers No. MS86-164* (1986), p. 1.

87. Zisk, B.I., "The Appeal of Flexible Manufacturing," *CIM Review* Fall 1984, p.67

88. Zisk, B.I., "The Appeal of Flexible Manufacturing," *CIM Review* Fall 1984, p.71

89. Zygmont, J., "Flexible Manufacturing Systems: Curing the Cure-All," *High Technology*, October 1986, pp.22-23

INDEX

Abernathy 138
Afentakis 19, 21, 138
Analytical Models of FMS 14
Annualized Inventory Holding Cost 112
Askin 22, 138
Assumptions 25-27, 33, 48, 128
Batch flow 29, 41
Biegel 13, 144
Blynsky 138
Bottleneck work center 59, 60, 62, 64, 69, 72, 79, 82, 86, 95-97, 100, 101, 109, 114, 125, 128, 129
Brown 138
Buffa 139
Burstein 16, 139
Buzacott 14, 15, 29, 139
Bylinsk 139
Capacity Utilization 21, 31, 47, 48, 67-69, 71, 72, 74, 75, 76, 85, 89, 90, 94-96, 98, 108, 111-113, 117, 118, 121-123, 125, 126, 130
Cellular Manufacturing 5, 16, 19, 147
Chakravaty 139
Chang 18, 139
Choong 20, 144
Church 5, 26, 139
CIMS 5, 17, 23
Cincinnati Milacron 4
Cmin 36, 47, 51, 64, 69, 79, 86, 91, 101, 109, 111, 114
Co 26, 140, 146, 147
Computer Integrated Manufacturing System 5
Configuration I 8, 62-65, 67-69, 71, 72, 74, 76, 77, 78-80, 82-84, 96, 103, 126, 127, 132
Configuration II 8, 60, 100, 103-108, 110, 111, 113, 114-127, 134

Configuration III 8, 79, 81-85, 87, 89, 90, 92-96,
        98, 99, 102, 103, 119, 126, 127,
        128, 129, 135
Conventional systems 11, 13, 18, 20, 23, 26, 27,
        51, 52
Costa 16, 32, 140
Cudworth 140
Cutosky 16, 140
Dannenbring 140
Dar-El 26, 140
Definitions 29
Delporte 24, 140
Economic Production Quantity 22, 23, 146
Enscore 25, 144
EOQ 22
EPQ 22, 23
Factory automation 3, 4, 7, 11, 12, 14, 29, 128
Ferrario 14, 140
Flexible Assembly Systems 17
Flexible Machining Systems 18
Flexible Manufacturing Island 6, 9, 29
Flynn 16, 140
FMS 3-7, 11-20, 23, 24, 26-29, 131, 138, 141-147
Galvin 25, 141
Goldhar 146
Gomma 141
Graves 20, 23, 141
Greshwin 141
Group technology 16, 19, 140
Hall 17, 141
Hartley 3, 141
Harvey 12, 141
Hassan 146
Hayes 142
Hierarchial Approach 16
Hildebrant 13, 18, 142, 146
Hill 142
Hughes 3, 142
Hyer 5, 147

Index 149

Implementation of FMS 12
Inventory Holding Cost 6-9, 21, 25, 26, 30, 31, 34, 36, 37, 41-44, 47, 50, 52, 55, 56, 58, 59, 61, 66-68, 71, 72, 74, 76, 77, 78, 83-85, 87, 89, 91-99, 104, 105-108, 110-117, 119-130
Islands of automation 4, 5
Item sequence 8, 25, 34, 37, 38, 42-44, 47, 51, 59, 61, 64, 67, 70, 73, 82, 83, 86, 101, 104-107, 109, 112, 115-117, 131
Item sequences 8, 22, 24, 27, 33, 37, 38, 42-44, 61, 66, 67, 70, 84, 92, 104, 109, 110, 111, 114, 115, 119
Item sequencing 24, 26, 27
Jain 28, 142
Kaplan 21, 30, 142
Karmarkar 20-22, 30, 138, 142, 143
Karmarkar, U 142, 143
Karmarkar, U.S. 142, 143
Kilbridge 26, 147
Kim 20, 145
Kiran 143
Klahorst 143
Kusiak 20, 138, 143, 146
Leachman 25, 143
Lead time 13, 21, 22
Line-balancing 24
Litteral 143
Lot sizing 17, 18, 20-23, 27, 138, 140, 143, 145
Mabert 19, 20, 36, 143
Maimon 16, 20, 144
Makespan 25
Malone 144
Manufacturing Cycle time 22, 23, 146
Mastratta 146
Mathematical models 14, 15, 27
Mean-Value-Analysis-Que 13
Methodology 17, 23, 25, 28, 29, 34, 38, 44, 58, 119, 128

Minimum feasible production cycle duration 36
Mize 4, 144
Morrion 145
Multiple-product, Multiple-Machine studies 25
Multiple-products 24
Nawaz 25, 144
Office of Technology Assessment 11
Phillips 28, 145
Pinto 19, 20, 36, 143
Piszczalski 14, 144
Production cycle duration 8, 22, 31, 33, 34, 36, 48, 129
Programmable Automation 11
Ranky 29, 144
Rate of Increase in Holding Cost 31, 47, 96, 97, 107, 125, 130
Ritzman 25, 145
Salomon 144
Seifiet 4, 144
Selinger 145
Sensitivity Analysis 32, 47, 49, 50, 52, 58, 60, 67, 70, 73, 74, 84, 85, 88, 93, 96, 107, 111, 112, 117, 120, 125, 130, 132, 133-136
Sequence-dependent setup 25
Sequencing studies 26
Setup Time 3, 16, 19, 29-31, 35, 38, 47, 49, 65, 67, 79, 96, 114, 125, 130
Shannon 28, 145
Shin 20, 145
Single-stage 23
Skeith 23, 146
Smith-Daniel 25, 145
Spur 145
Stecke 14, 17, 20, 138, 139, 141-145
Stochastic demand 22, 25, 143
Stochastic scheduling 20, 141
Suri 12, 13, 15, 145, 146
System Capacity Utilization 31, 47, 48

System utilization 6
Szendrovits 22, 23, 146
Taha 23, 146
Talaysum 146
Tersine 23, 146
Thomopolous 26, 146, 147
Unit flow 29, 33, 51, 55, 69, 72, 91, 114
Vachajitpan 25, 146
Vaithianathan 19, 146
Valcada 15, 146
Venkitaswamy 147
Viehweger 145
Wagner 20, 147
Wemmerlov 5, 147
Wester 26, 147
Whitney 12, 15, 140, 145
Wilhelm 15, 20, 28, 145, 147
WIP Inventory 22
Within 6, 12, 29, 43, 59, 96, 125, 128, 130, 147
Wortman 28, 147
Zisk 3, 147
Zygmont 147